Collins Revision

NEW GCSE

MATHS

Higher

Revision Guide

For **GCSE Maths from 2010**

Edexcel + **AQA** + **OCR**

Keith Gordon

Published by Collins
An imprint of HarperCollins*Publishers*
77–85 Fulham Palace Road
Hammersmith
London W6 8JB

Browse the complete Collins catalogue at
www.collinseducation.com

10 9 8 7 6

ISBN 978-0-00-734099-6

British Library Cataloguing in Publication Data
A catalogue record for this publication is available from the British Library.

Written by Keith Gordon
Edited by Christine Vaughan and Marie Taylor
Project Managed by Philippa Boxer
Design by Graham Brasnett
Illustrations by Kathy Baxendale and Gray Publishing
Index compiled by Marie Lorimer
Printed and bound in China

Collins Revision

NEW GCSE

MATHS

Higher

Revision Guide

Keith Gordon

Contents

If you were in Year 10 or below in September 2010, you will be taking the new GCSE examination. The first one is in June 2012 but in fact, because all the examination boards are offering modular courses, you could be taking a modular examination as early as November 2010!

Each examination board has a slightly different way of organising their modules and different content is assessed in each of them, so you need to ask your teachers what examination you will be doing. There are three examination boards in England: AQA, EDEXCEL and OCR.

Once you know the examination you will be taking you can identify the sections to revise for any module by looking at top of each page in this revision guide: the modules each topic is assessed in by each board is listed. You will see that some of the topics appear in more than one module. This is because some topics in mathematics – such as basic number and basic algebra – are needed for all aspects of mathematics. Your teachers will be able to give you more guidance on the specific topics to be revised for each modular examination.

Remember, this is a revision guide and it is not intended to teach you any mathematics. Use it to remind you of the mathematics you have already learned. It will also give you some useful tips about the ways that certain topics are assessed in the examination.

To get the most from your revision, follow these steps:

- Read each page carefully.
- Work through the examples, checking that you understand and agree with each step.
- Read the 'Remember' bubbles – these give you advice on how to prepare for the examination.
- Work through the questions at the bottom of each page.
- Check your answers.
- If you are confident that you have mastered the topic, try the questions in the workbook.
- Check the answers to the workbook questions – have you have given the full answer that the examiners are looking for?

Each page of this revision guide is structured the same way.

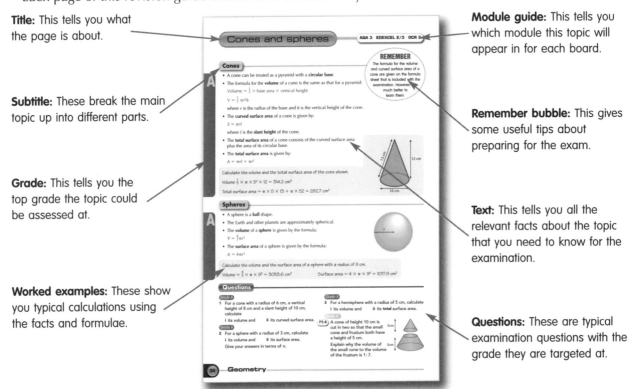

Title: This tells you what the page is about.

Subtitle: These break the main topic up into different parts.

Grade: This tells you the top grade the topic could be assessed at.

Worked examples: These show you typical calculations using the facts and formulae.

Module guide: This tells you which module this topic will appear in for each board.

Remember bubble: This gives some useful tips about preparing for the exam.

Text: This tells you all the relevant facts about the topic that you need to know for the examination.

Questions: These are typical examination questions with the grade they are targeted at.

The revision guide has corresponding write-in workbook pages and the answers are in a detachable section at the back. The workbook also gives the grades of questions. When you check your answers, you will see that the mark scheme breaks the marks down so that you can see if you showed the necessary working.

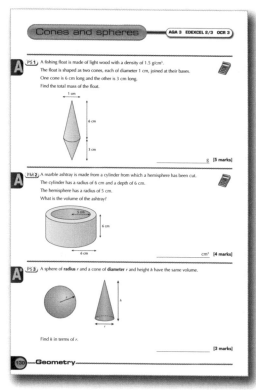

From September 2010 there will be a new mathematics GCSE. The first certificates will be awarded in June 2012.

For the English Awarding Bodies (AB) there are six specifications available.

AQA Mathematics Linear

This is a linear course – you will take two papers, one non-calculator and one calculator, in the same examination session. Each paper assesses all the topics in the specification, which means that you will get questions on Number and Algebra, Geometry and Measures, and Statistics and Probability in both papers.

AQA Mathematics Modular

This is a three-unit modular course.

Unit 1 is Statistics and Number

Unit 2 is Number and Algebra

Unit 3 is Geometry and Algebra

EDEXCEL Specification A

This is a linear course – you will take two papers, one non-calculator and one calculator, in the same examination session. Each paper assesses all the topics in the specification, which means that you will get questions on Number, Algebra, Geometry, Measures, Probability and Statistics in both papers.

EDEXCEL Specification B

This is a three-unit modular course.

Unit 1 is Statistics and Probability

Unit 2 is Number, Algebra, Geometry 1

Unit 3 is Number, Algebra, Geometry 2

OCR Specification A

This is a linear course – you will take two papers, one non-calculator and one calculator, in the same examination session. Each paper assesses all the topics in the specification, which means that you will get questions on Statistics, Number, Geometry and Measures, and Algebra in both papers.

OCR Specification B

This is a three-unit modular course. All three units cover Statistics, Number, Algebra, and Geometry and Measures.

Find out which course you are taking

Your teacher will be able to tell you which topics are assessed in each module, or you can find out from the examination boards' websites.

If you are taking a linear course, you will need to revise all the topics in this book to prepare for the final examination. If you are taking a modular course you will need to revise only the topics in each module before each examination. For example, you might do Module 1 in November of Year 10, Module 2 in June of Year 10 and Module 3 in June of Year 11. Look at the top of the pages in this book to see which modules a topic is assessed in.

What's new?

The new GCSE has some new types of question that haven't appeared in the GCSE examination before.

Functional mathematics

This is all about 'real-life' mathematics. They are not flagged on the examination paper, but in this revision guide and workbook functional questions are labelled with (FM). One way of deciding if a question is functional is to decide if anyone would want to know the information in a real-life context.

For example:
'Work out 30% of £150' is not functional but 'Work out the new price of a suit normally costing £150 after a 30% reduction in a sale' is a functional question.

In the Foundation Paper, 40% of the questions will have a functional mathematics element. In the Higher paper, 20% will have a functional element. Many functional mathematics questions will be very familiar to you (you've probably come across questions like the one above about the suit reduced in a sale in past papers) but some may be new. For example, unless you have done a Functional Mathematics Certificate, you may not have come across this sort of question before:

Example (Grade D)

Julie belongs to a local gym.

This table shows the calories used on four types of exercise machines at the gym.

Exercise machine	Number of calories per minute		
	Intensity		
	Low	Med	High
Rowing machine	8	11	14
Treadmill	6	9	12
Static bicycle	5	8	11
Cross-trainer	10	14	18

Julie goes to the gym for an hour.

Julie likes to do about 30 minutes of low-intensity exercise and no more than 10 minutes of high-intensity exercise.

Devise a training programme that uses each machine and allows Julie to use a total of at least 500 calories.   **[5 marks]**

Answer

This is an **open-ended** question – it does not have a definite answer.

It is important to make sure that you communicate your answer to the examiner and do enough mathematics to get all the marks available.

This would be a good answer:

15 minutes on the rowing machine at low intensity will be 15 × 8 = 120 calories used.

10 minutes on the cross-trainer at high intensity will be 10 × 18 = 180 calories used.

20 minutes on the static bicycle at medium intensity will be 20 × 8 = 160 calories used.

15 minutes on treadmill at low intensity will be 15 × 6 = 90 calories used.

Total calories = 120 + 180 + 160 + 90 = 550 calories.

There are many other possible answers: the examiners will not be looking for a single 'right' answer – instead they will be checking that all the conditions are met and that the calculations are correct.

For example:

Are all machines used? ✔

10 minutes at high intensity? ✔

30 minutes at low intensity? ✔

1 hour in total? ✔

At least 500 calories? ✔

All calculations correct? ✔

Assessment Objective 1 (AO1)

About half of the questions in the examination will be straightforward questions that test if you can do mathematics. These are known as Assessment Objective 1 (AO1) questions. The examples below are AO1 questions, with advice on how to answer them. You will have come across this sort of question often.

Example (Grade D)

In a school there are 600 students and 30 teachers.

12% of the students are vegetarian.

10% of the teachers are vegetarian.

How many vegetarians are there in the school altogether?   **[3 marks]**

Answer

$0.12 \times 600 = 72$ students are vegetarian | Use the percentage multipliers 0.12 and 0.1 to work out 12% and 10%. |

$0.1 \times 30 = 3$ teachers are vegetarian

$3 + 72 = 75$ total vegetarians | Show the total clearly. |

Make sure that all calculations are shown. It is always a good idea to use percentage multipliers in this sort of question as it makes calculations easier.

Example (Grade D)

Solve the equation $6x + 5 = 9 - 2x$ **[3 marks]**

Answer

$6x + 5 = 9 - 2x$

$6x + 2x = 9 - 5$

$8x = 4$

$x = 0.5$

Assessment Objective 2 (AO2)

About 30% of the examination questions are designed to test whether you understand the topics and can apply mathematics in slightly more involved situations. These are known as Assessment Objective 2 (AO2) questions. They are similar to the old 'Using and Applying Mathematics' questions, except that they are designed to test your **understanding**, not just whether you can do the mathematics. They are not flagged in the examination, but in this revision guide and workbook AO2 questions are labelled with (AU). The examples opposite are AO2 questions, with advice on how to answer them.

Example (Grade E)

Here are the instructions for cooking a joint of beef.

Cook for 20 minutes at 200 °C.

Reduce the temperature to 160 °C and cook for 10 minutes per pound.

Kevin has a 6 pound joint of beef that he wants to take out of the oven at 1 pm.

What time should he start to cook it?

[3 marks]

Answer

Time to cook $= 20 + 6 \times 10$

$= 80$ minutes

> Show the calculation for working out the cooking time. This will get a method mark. The time of 80 minutes will get an accuracy mark.

Time to put in oven is 80 minutes before 1pm

$= 11.40$ am

> Work out what time is 80 minutes before 1 pm. This gets 1 accuracy mark.

Example (Grade D)

An isosceles triangle has one angle of 70°.

Work out what the other angles could be.

[3 marks]

Answer

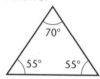

> Drawing a diagram will get full marks as will writing out the values.

The other possible triangle has angles of 70°, 55° and 55°.

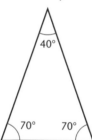

> The question doesn't tell you which angle is 70°. It could be the equal angle or it could be the non-equal angle, so show both possibilities.

The other possible triangle has angles of 40°, 70° and 70°.

Assessment Objective 3 (AO3)

About 20% of the examination questions are designed to test whether you can solve problems. These are known as Assessment Objective 3 (AO3) questions. They are not flagged in the examination, but in this revision guide and workbook AO3 questions are labelled with (PS). The following examples are AO3 questions, with advice on how to answer them.

Example (Grade C)

At the school summer fair, Ray sets up a stall to raise money.

Players pay 50p to throw two dice.

If they throw two 6s they get £2 back. If they throw one 6 they get £1 back.

If 100 people play the game, how much money would Ray expect to raise?

You may use the grid on the next page to help you.

[4 marks]

		Score on first dice					
	+	1	2	3	4	5	6
Score on second dice	1						
	2						
	3						
	4						
	5						
	6						

Answer

P(two 6s) $= \frac{1}{36}$

P(one 6) $= \frac{10}{36}$

Use the table to work out the probabilities.

Number of two 6s in 100 throws $= \frac{1}{36} \times 100 \approx 3$ which is £6 paid out

Number of one 6 in 100 throws $= \frac{10}{36} \times 100 \approx 28$ which is £28 paid out

Use the probabilities to work out the expected values.

Income = £50

Profit = £50 − £28 − £6 = £16

Work out the expected profit.

Example (Grade D)

Tammy makes three-legged stools and four-legged tables.

The stools sell for £12 and the tables sell for £30.

One day she sells enough stools and tables to make £180 and uses 31 legs.

How many stools and tables did she sell? **[3 marks]**

Answer

1 table costs £30 which leaves £150 which does not divide by 12.

Assume one table is sold, which does not work.

2 tables cost £60 which leaves £120 which would be 10 stools.

Assume two tables are sold, which leaves 10 stools.

This would be $2 \times 4 + 10 \times 3 = 38$ legs.

Work out how many legs this is.

3 tables cost £90 which leaves £90 which does not divide by 12.

Keep on increasing the number of tables.

4 tables cost £120 which leaves £60 which would be 5 stools.

This would be $4 \times 4 + 3 \times 5 = 31$ legs.

Eventually the number of table and stools will give the correct number of legs.

Quality of written communication

Some questions in the examination are designed to assess how well you set out and explain your answers, or your Quality of Written Communication (QWC). All examination boards will indicate in some way which questions are allocated marks for QWC.

This is another new requirement for mathematics, although it has been assessed in other GCSE subjects for some time.

To earn QWC marks, you will need to:

- write legibly, with accurate use of spelling, grammar and punctuation in order to make your meaning clear

- select and use a form and style of writing appropriate to purpose and to complex subject matter

- organise relevant information clearly and coherently, using specialist vocabulary when appropriate.

To help you do this, think about whether what you have written will make sense to someone else. Make sure your answer is clear and logical. This is a good idea anyway, because if the examiner can understand your method, you might get some marks.

You should also make sure that you use correct mathematical notations, such as £3.60 not £3.6, or $5x$ not $x5$, and use specialist vocabulary: for example, say 'because it is an alternate angle' rather than 'it is a Z angle'.

Here is an example of a question that has a QWC mark allocated to it.

Example (Grade E)

This bar chart shows the number of boys and girls who have a salad for school lunch in one week.

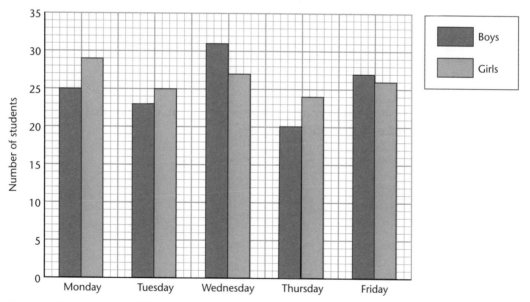

The headteacher says 'More girls than boys have salad for lunch'.

Is the headteacher correct?

Justify your answer.

[3 marks]

Mark scheme

Answer	Mark
Working out the total for boys or girls	M1
Correct totals for boys (126) **and** girls (131)	A1
Valid conclusion using data e.g. Head is correct as 131 > 126	QWC1

Answer

Here are three answers.

126

131

Yes 131 > 126

> This would just about get three marks as it has all the necessary information, but there is not much communication in this answer.

Total for boys = 136

Total for girls = 131

The head is wrong as more boys have salad as 136 > 131

> This answer would get two marks out of three (M1, A0, QWC1). Even though the total for the boys is wrong, the total for the girls is correct, which earns the method mark (M1). The conclusion is correct for the values calculated, earning the communication mark (QWC1). However, the total for boys is wrong, so the accuracy mark is zero (A0).

Total for boys = 136

Total for girls = 121

The head is wrong as more boys have salad as 136 > 121

> This answer would get 1 mark out of three (M0, A0, QWC1) as both totals are wrong, but the conclusion is correct for the values calculated.

This shows that it is important to use words to communicate what you are doing so that you get the mark for communication even if you make mathematical mistakes.

Statistics 1

Averages and range

- The **mode** is the **most common** value in a set of data.

- The **median** is the **middle number** when the data items are arranged in order.

- The **mean** of a set of data is the sum of all the values in the set divided by the total number of values in the set.

- The mean is defined as:

$$\text{mean} = \frac{\text{sum of all values}}{\text{total number of values}}$$

- When people talk about 'the average', they are usually referring to the mean.

- The **range** of a set of data is the difference between the highest and lowest values in the set.

  range = highest value − lowest value

Frequency tables

- When a lot of data has to be represented, it can be put into a **frequency table**.

This table shows how many times students in a form were late in a week.

Number of times late	0	1	2	3	4	5
Frequency	11	8	3	3	3	2

> **REMEMBER**
> The mode is the value that has the highest frequency; it is *not* the frequency itself.

- Find the **mode** by looking for the data value that has the **highest frequency**.

- Find the **median** by adding up the frequencies of the data items, in order, until the half-way point of all the data in the set is passed.

- Find the **mean** by multiplying the value of each data item by its frequency, adding the totals, then dividing by the total of all the frequencies.

> **REMEMBER**
> In examinations, data is usually given in vertical tables. Add another column to work out $x \times f$ (value multiplied by the frequency).

To find the average number of times students were late from the table above, work out the total number of times students were late as:
$0 \times 11 + 1 \times 8 + 2 \times 3 + 3 \times 3 + 4 \times 3 + 5 \times 2$
and work out the total frequency as: $11 + 8 + 3 + 3 + 3 + 2$,
then divide the first total by the second.

Questions

Grade D

1 Find the mean, mode and median of these sets of data.

a 8, 9, 4, 7, 8, 4, 9, 6, 3, 8

b 11, 12, 10, 12, 15, 13, 11, 10, 11, 13, 14

Grade C

PS 2 Find five numbers that have a mode of 4, a median of 5, a mean of 6 and a range of 7.

☐ ☐ ☐ ☐ ☐

Grade D

FM 3 The number of times students in a form were late for school in a week is shown in the table above.

a How many students were there altogether in the form?

b What is the modal number of times students were late?

c What is the median number of times students were late?

d i What is the total number of times students were late?

ii What is the mean number of times students were late per week?

Statistics

Grouped data and frequency diagrams

- When there is a wide range of data, with lots of values, there are often too many entries for a frequency table. In this case, use a **grouped frequency table**.
- In a grouped frequency table, data is **recorded in groups** such as $10 < x \leqslant 20$.
- The notation $10 < x \leqslant 20$ means values between 10 and 20, not including 10 but including 20.
- Grouped data can be shown in a **frequency polygon**.
- In a frequency polygon, the **midpoint** of each group is plotted against the **frequency**.

This table shows the marks in a mathematics examination for 50 students. The frequency polygon shows the data.

Marks, x	Frequency, f
$0 < x \leqslant 10$	4
$10 < x \leqslant 20$	9
$20 < x \leqslant 30$	17
$30 < x \leqslant 40$	13
$40 < x \leqslant 50$	7

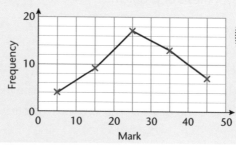

REMEMBER

If asked to calculate an estimate for the mean, add two columns to the table: the first for the midpoint, m, and the second for $m \times f$.

- The **modal class** is the group with the greatest frequency.
- The **median** cannot be found from a grouped frequency table.
- Calculate an **estimate of the mean** by adding the midpoints multiplied by the frequencies and dividing by the total frequency.

Histograms

- A **histogram** is similar to a bar chart, but is for continuous data only, such as time or weight.
- The **horizontal axis** has a **continuous** scale and there are no gaps between the bars.
- The **area** of each bar **represents** the **frequency** of the bar.
- The height of the bars is called the **frequency density** and is calculated by:

$$\text{frequency density} = \frac{\text{frequency of class interval}}{\text{width of class interval}}$$

This histogram shows the masses of 100 snails.

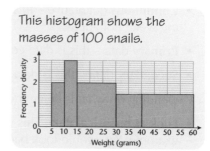

- The median is the value where the area on both sides is equal.
- The lower quartile is the value that splits the area in the ratio 1:3.
- The upper quartile is the value that splits the area in the ratio 3:1.

Questions

Grade C

1 The marks for 50 students in a mathematics examination are shown in the table above.
   a What is the modal class?
   b i What is the total of the 'midpoints × frequencies'?
     ii What is the estimated mean mark for the form?

Grade A

2 Refer to the histogram above.
   a Work out how many snails had a mass between 5 and 10 grams.

b Explain why the lower quartile is 15 grams.
c What is the median mass of the snails?

AU 3 Refer to the histogram above.
Decide whether each of the following statements is definitely true, definitely false or could be true. Explain your choice of answer.
   a The range of the weight of the snails is 50 g.
   b The number of snails between 30 and 40 grams is the same as the number of snails between 40 and 60 grams.
   c The approximate mean weight of the snails is 29.6 g.

Statistics

The data handling cycle

There are **four** parts to the data handling cycle.

- **Outlining** the problem and **planning**. This usually involves stating a **hypothesis**, which is the idea being tested.

- Stating how the data will be **collected**.

- Saying how the data will be **processed** and **represented**. This will usually involve working out means and measures of spread and showing the data in a suitable diagram such as a histogram or a box plot.

- **Interpreting** the results and making a **conclusion**. This should be related to the original hypothesis or problem.

> **REMEMBER**
> When you are asked to describe the data handling cycle, make sure you describe all four parts. Marks will be allocated for the overall quality and logic of your answer.

A gardener grows courgettes outside and in a greenhouse.

These are the lengths of 10 courgettes grown outside.

12   14   11   15   17   9   16   12   11   18

Ten courgettes from the greenhouse also have their lengths measured.

The mean length is 15.5 cm. The range of the lengths is 6 cm.

Investigate the hypothesis:

'Courgettes grown in the greenhouse are bigger than those grown outside.'

Data and data collection

There are different types of data:

- **Qualitative** – this is data such as the colour of cars.

- **Discrete** quantative – this is numerical data that can only take certain values, such as the number of people on a bus.

- **Continuous** quantative – this is numerical data that can take any value in a range of values, such as the weight of apples.

- **Primary** data – this is data collected using **surveys** or **experiments**.

- **Secondary** data – this is data collected from existing **lists** or **tables**.

Depending on what is being investigated, there are four different ways to collect data:

- **Look at historical data**. This would be used to predict whether it will rain on 4 July in Edinburgh, for example. You would look back at the records for all previous years.

- **Experiment**. This would be used to test the probability of an event that cannot be predicted, such as how many blue balls are in a bag. By taking out 100 balls, you can get an approximate probability.

- **Equally likely outcomes**. This is used when the probability of an event can be predicted. For example, taking a king from a pack of cards.

- **Surveys**. These are used when there is no historical data and no way to predict probabilities. A survey, for example, would be used to find out what subject is the favourite amongst Year 7 pupils.

Questions

Grade C

FM 1 Investigate the hypothesis that 'Courgettes grown in a greenhouse are bigger than those grown outside' using the data above.

Grade B

FM 2 Describe how a headteacher can use the data handling cycle to investigate the hypothesis 'Girls do better in GCSE mathematics examinations than boys'.

Surveys

- **A survey** is an organised way of finding people's opinions or testing an hypothesis.

- Data from a survey is usually collected on a **data collection sheet**.

- Questionnaires are used to collect a lot of data.

- Questions on questionnaires should follow some rules.
  - Never ask a leading question
  - Never ask a personal question
  - Keep questions simple, with a few responses
  - Make sure the responses cover all possibilities and do not include any overlapping responses

> **REMEMBER**
> When you are asked to criticise questions in an examination, keep your answers short and simple, for example, 'It is a leading question'.

Sampling

- Statisticians often have to **collect information** or **test hypotheses** about a **population**.

- In statistics, a **population** can be **any group** of people, objects or events.

- **Sampling** is a method of finding information, using a small sample from a large population.

- A sample needs to be:
  - **representative**: covering all the different groups within a population without bias
  - **of sufficient size**: large enough to make sure the results are valid for the whole population

- There are two main types of sample:
  - **random**: all members of the population have an equal chance of being chosen
  - **stratified**: the population is divided into categories and a number of each category is surveyed in the same proportion as in the population. The sample within each category is taken randomly.

> **REMEMBER**
> When asked to comment on sampling methods in an examination, keep your answers short and simple such as, 'It is not a random process'.

This table shows the numbers of students in each year of a school.

School year	Boys	Girls	Total
7	52	68	120
8	46	51	97
9	62	59	121
10	47	61	108
11	39	55	94

Questions

Grade D

FM 1 The following are two questions used in a survey about recycling.

Give one reason why each question is not a good one.

a Recycling is a waste of time and does not help the environment.
Don't you agree? ☐ Yes   ☐ No

b How many times a month do you use a bottle bank?
☐ Never   ☐ 2 times or less
☐ More than 4 times

Grade A

FM 2 a The table above shows the numbers of students in each year of a school. The headteacher wants to survey 50 of these students. Which method below is the best? Give a reason why the other two are not suitable.

i Asking two year 7 forms of 25 students

ii Asking 50 students on the first bus in the morning

iii Put the names of all students in a hat and pick 50 names.

b How many students from each year group should the head pick to get a 10% stratified sample?

Statistics 2

Line graphs

- **Line graphs** are used to show how data changes over a period of time.

- Line graphs can be used to show **trends**, such as how the average daily temperature changes over the year.

- **Data points** on line graphs can be joined by **lines**.

  – When the lines join points that show **continuous data**, such as points showing the height of a plant each day over a week, they are drawn as **solid lines** because the lines can be used to estimate intermediate values.

  – When the lines join points that show discrete data, such as the average daily temperature, they are drawn as **dotted lines**, because the lines cannot be used to estimate intermediate values.

REMEMBER
The largest increase or decrease on a line graph is at the steepest part of the graph.

This line graph shows the temperatures in a town in Australia and a town in Britain.

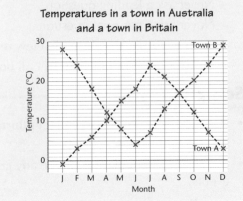

Temperatures in a town in Australia and a town in Britain

Stem-and-leaf diagrams

- When data is first recorded, it is called **raw data** and is **unordered**.

- Unordered data can be put into order to make it easier to read and understand. This is called **ordered** data.

- A **stem-and-leaf diagram** is a way of showing ordered data.

- The **stem** is the number on the left of the vertical line. The **leaves** are the numbers on the right of the vertical line.

- The **mode** is the most common entry.

- You can find the **median** by counting from the start to the middle value.

- You can find the **range** by subtracting the lowest value from the highest value.

- You can find the **mean** by adding all the values and dividing by the total number of values in the table.

REMEMBER
Always make sure you add a key to show what the stem and leaves represent.

This stem-and-leaf diagram shows the marks scored by students in a test.

```
1 | 3 4 5 6
2 | 0 1 1 3 4 5 8
3 | 1 2 5 5 5 9
4 | 2 2 9
```

Key 1 | 3 represents 13 marks

Questions

Grade D

FM 1 Refer to the line graph above.

  a Which town is hotter, on average? Give a reason for your answer.

  b Which town is in Australia? Give a reason for your answer.

  c In which month was the average temperature the same in both towns?

  d Is it true that the average temperature was the same in both towns on a day in early April?

Grade D

2 Refer to the stem-and-leaf diagram above.

  a How many students took the test?

  b What is the range of the marks? Show your working.

  c What is the modal mark?

  d What is the median mark?

  e What is the mean mark? Show your working.

Scatter diagrams

Scatter diagrams

- A **scatter diagram** (also known as a **scattergraph** or **scattergram**) is a diagram for comparing two variables.
- The variables are plotted as **coordinates**.

Here are 10 students' marks for two tests.

Tables	3	7	8	4	6	3	9	10	8	6
Spelling	4	6	7	5	5	3	10	10	9	7

This is the scatter diagram for these marks.

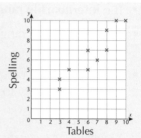

Correlation

- The scatter diagram will show a **relationship** between the variables if there is one.
- The relationship is described as **correlation** and can be written as a '**real-life**' statement.

Positive correlation

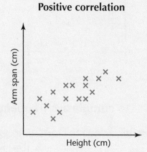

For the first diagram: 'The taller people are, the bigger their arm span'.

Negative correlation

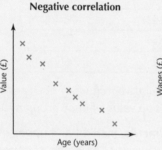

No correlation

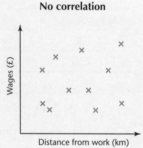

Line of best fit

- A **line of best fit** can be drawn through the data.
- The line of best fit can be used to **predict** the value of one variable when the other is known.

REMEMBER

Draw the line of best fit between the points with about the same number of points on either side of it.

The line of best fit passes through the 'middle' of the data.

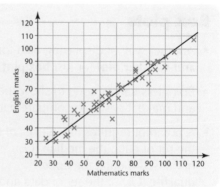

Questions

FM 1 **a** Refer to the scatter diagram of students' scores for tables and spelling tests, above. What type of correlation does the scatter diagram show?

b Describe the relationship in words.

FM 2 **a** Describe, in words, the relationship between the value of a car and the age of the car, shown in the second of the three scatter diagrams above.

b Describe, in words, the relationship between the wages and the distance travelled to work, shown in the third of the three scatter diagrams above.

FM 3 Refer to the line of best fit on the scatter diagram showing the English and mathematics marks. Estimate the score in the English examination for someone who scored 75 in the mathematics examination.

Statistics

Cumulative frequency and box plots

Cumulative frequency diagrams

- The **interquartile range** is a measure of the **dispersion** of a set of data.

- The interquartile range **eliminates extreme values** and bases the measure of spread on the middle 50% of the data.

- The interquartile range and the median can be found by drawing a cumulative frequency diagram. Add the frequency to the sum of the preceding frequencies to find the cumulative frequency.

This grouped table shows the marks of 50 students in a mathematics test and the cumulative frequencies.

The graph shows the cumulative frequency diagram for the same data.

Mark	Number of students	Cumulative frequency
21 to 40	7	7
41 to 60	14	21
61 to 80	14	35
81 to 100	13	48
101 to 120	2	50

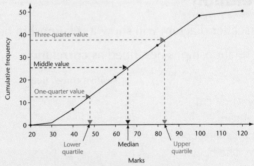

- The points of a cumulative frequency diagram are plotted as the **top value** of each group against its **cumulative frequency**.

- To find the **lower quartile**, **median** and **upper quartile**, draw horizontal lines from the quarter, half and three-quarter values to the graph, then read the values from the horizontal axis.

- To find the **interquartile range**, subtract the lower quartile from the upper quartile.

Box plots

> **REMEMBER**
> Box plots are always plotted against a **scale** so that their values can be accurately read and plotted.

- Another way of displaying data for comparison is by means of a **box-and-whisker plot** (or just **box plot**).

- Box plots can be used to show the spread of two or more sets of data and make it easier to compare data.

- A box plot requires five pieces of data: the **lowest value** of the set of data, the **lower quartile** (Q_1) of the set of data, the **median** (Q_2) of the set of data, the upper **quartile** (Q_3) of the set of data and the **highest value** of the set of data.

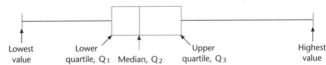

Questions

FM 1 Refer to the cumulative frequency diagram above.

  a What is the median score?

  b What is the lower quartile?

  c What is the upper quartile?

  d What is the interquartile range?

  e The lowest score is 21. The highest score is 120. Draw a box plot to show the data using a scale from 20 to 120.

AU 2 How could you tell from a box plot that it shows a symmetrical distribution?

Statistics

Probability

Relative frequency

- The probability of an event can be found by carrying out many trials and calculating the **relative frequency** or **experimental probability**.

- To calculate the relative frequency of an event, **divide** the number of times the event occurred during the experiment by the **total number of trials** done in the experiment.

> **REMEMBER**
>
> If you are asked who has the best set of results, always say the person with the most trials.

- The **higher the number of trials** carried out, the nearer the experimental probability will be to the true probability.

These are the results when three students tested the spinner.

Student	Ali	Barry	Clarrie
Number of throws	20	60	240
Number of 4s	5	13	45

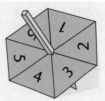

Mutually exclusive and exhaustive events

- When the outcome of event A can never happen at the same time as the outcome of event B, then event A and event B are said to be **mutually exclusive**.

> If a bag contains black and yellow balls, the events 'picking a yellow ball' and 'picking a black ball' can never happen at the same time when only one ball is taken out: that is, a ball can be either black or yellow.

- When the probability of mutually exclusive events add up to 1, they are called **exhaustive events**.

> Throwing an odd or an even number when throwing a dice is an exhaustive event because you will either throw an odd number or an even number.

- If there is an event A, the **complementary** event of A is event A not happening.

> The probability of drawing an ace from a pack of cards is $\frac{4}{52} = \frac{1}{13}$, so the probability of not drawing an ace is $1 - \frac{1}{13} = \frac{12}{13}$.

Questions

Grade C

1. The table above shows the results when three students tested a home-made spinner.

   a For each student, calculate the relative frequency of getting a 4. Give your answers to 2 decimal places.

   b Which student has the most reliable estimate of the actual probability of a 4? Give a reason for your answer.

   c If the spinner was fair, how many times would you expect it to land on 4 in 240 spins?

Grade B

2. a A card is taken at random from a pack of cards.

   i What is the probability it is red?

   ii What is the probability it is black?

   iii Explain why the events in parts **i** and **ii** are mutually exclusive.

   iv Explain why the events in parts **i** and **ii** are exhaustive.

   b A card is taken at random from a pack of cards.

   i What is the probability it is a king?

   ii What is the probability it is a **not** a king?

Statistics

Expectation

- When the probability of an event is known, you can predict how many times the event is likely to happen in a given number of trials.

- This is the **expectation**. It is *not* what is going to happen.

> If a coin were tossed 1000 times, the expectation would be 500 heads and 500 tails. It is very unlikely that this would actually occur in real life.

- The expected number is calculated as: **expected number = P(event) × total trials**

Two-way tables

- A **two-way table** is a table that links two variables.

- One of the **variables** is shown by the rows of the table.

- One of the **variables** is shown by the **columns** of the table.

This table shows the nationalities of people on a plane and the types of tickets they have.

	First class	Business class	Economy
American	6	8	51
British	3	5	73
French	0	4	34
German	1	3	12

REMEMBER

Usually a column and a row for the totals are included in the table. If they are not, always add up each row and column anyway. You will almost always need these values to answer the questions.

Addition rule for events

- To work out the probability of either of two mutually exclusive events happening, **add** the probabilities of all the separate events: **P(A or B) = P(A) + P(B)**

Questions

Grade C

1 A bag contains ten counters. Five are red, three are blue and two are white. A counter is taken from the bag at random. The colour is noted and it is replaced in the bag. This is repeated 100 times.

   a How many times would you expect a red counter to be taken out?

   b How many times would you expect a white counter to be taken out?

   c How many times would you expect a red or a white counter to be taken out?

Grade D

FM 2 Refer to the table above.

   a How many travellers were on the plane altogether?

   b What percentage of the travellers had first-class tickets?

   c What percentage of the business-class passengers was American?

Grade D

3 A card is taken at random from a pack of cards.

   a What is the probability that it is an ace?

   b What is the probability that it is a king?

   c What is the probability that it is an ace or a king?

   d What is the probability that it is not an ace?

   e What is the probability that it is not an ace or a king?

Combined events

- When two events occur together they are known as **combined events**.
- The outcomes of the combined events can be shown as a list.

  If two coins are thrown together, the possible outcomes are (head, head), (head, tail), (tail, head) and (tail, tail).

- Another method for showing the outcomes of a combined event is to use a **sample space diagram**.

  If two dice are thrown, the outcomes can be shown as:

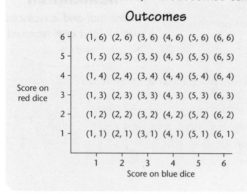

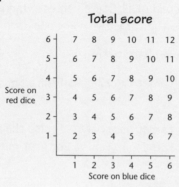

Tree diagrams

- An alternative method for showing the outcomes of combined events is a **tree diagram**.

- Note that the probabilities across any branch add up to 1.

- The probability of any outcome is calculated by multiplying together the probabilities on its branches.

  The probability of a head followed by a 6 has been calculated in the tree diagram on the right.

This tree diagram shows the outcomes of the combined events, tossing a coin followed by throwing a six with a dice.

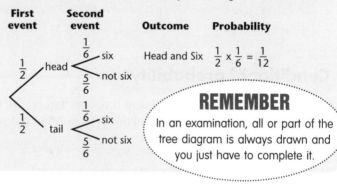

REMEMBER

In an examination, all or part of the tree diagram is always drawn and you just have to complete it.

Questions

Grade C

1 a When two dice are thrown together, how many possible outcomes are there?

b Refer to the left-hand sample space diagram for throwing two dice, above.

  i What is the probability of throwing a double with two dice?

  ii What is the probability that the difference between the scores on the two dice is 4?

c Refer to the right-hand sample space diagram for throwing two dice, above.

  i What is the probability of throwing a score of 5 with two dice?

ii What is the probability of throwing a score greater than 9 with two dice?

iii What is the most likely score with two dice?

Grade B

2 Using the tree diagram above, what is the probability of each of the following outcomes?

a a head followed by not a six

b a tail followed by a six

c a tail followed by not a six

Statistics

Independent events

- If the outcome of event A cannot affect the outcome of event B, then events A and B are known as **independent events**.

- Most of the combined events we have revised so far with sample space diagrams and tree diagrams have been independent events.

- An alternative method for working out the outcomes of combined independent events is to use **and** and **or**.

> On a fairground stall, Steven shoots darts at a target. The probability that he hits the target with any one dart is $\frac{2}{3}$. He has two darts.
>
> What is the probability that he:
>
> **a** hits the target twice?
> P(hits both times) = P(first dart hits **and** second dart hits)
> = P(hit) × P(hit) = $\frac{2}{3} \times \frac{2}{3} = \frac{4}{9}$.
>
> **b** hits the target once only?
> P(hits the target once only) = P(first hits **and** second misses **or** first misses **and** second hits) = $\frac{2}{3} \times \frac{1}{3} + \frac{1}{3} \times \frac{2}{3} = \frac{2}{9} + \frac{2}{9} = \frac{4}{9}$.

> **REMEMBER**
> Notice that **and** is replaced with × and **or** is replaced with +.

At least problems

- In examination questions concerning combined events, it is common to ask for the probability of at least one of the events occurring.

- As we are dealing with independent events, **P(at least one) = 1 − P(none)**.

> A fair dice is thrown three times. What is the probability of at least one 6 being thrown?
> P(at least one 6) = 1 − P(no 6s) = $1 - \frac{5}{6} \times \frac{5}{6} \times \frac{5}{6} = 1 - \frac{125}{216} = \frac{91}{216}$.

Conditional probability

- When the probability of an event is dependent on the outcome of another (preceding) event, it is known as conditional probability.

> A drawer contain seven socks, three are red and four are blue.
>
> Two socks are taken out of the drawer at random.
>
> What is the probability that they are the same colour?
> P(same colour) = P(red, red **or** blue, blue) = $\frac{3}{7} \times \frac{2}{6} + \frac{4}{7} \times \frac{3}{6} = \frac{3}{7}$.

> **REMEMBER**
> A tree diagram can be used to solve conditional probability problems.

Questions

Grade A

1 Annie throws a biased dice twice. The probability of throwing a six with the dice is $\frac{1}{4}$. What is the probability that Annie throws:

  a two sixes?

  b at least one six?

Grade A*

2 A box contains four white eggs and two brown eggs. Karen needs two eggs to bake a cake. She takes two eggs at random from the box. What is the probability that:

  a both eggs are brown?

  b both eggs are white?

  c one egg is white and one egg is brown?

Statistics grade booster

I can...

☐ draw an ordered stem-and-leaf diagram

☐ find the mean of a frequency table of discrete data

☐ draw a frequency polygon for discrete data

☐ find the mean from a stem-and-leaf diagram

☐ predict the expected number of outcomes of an event

☐ draw a line of best fit on a scatter diagram

☐ recognise the different types of correlation

☐ design a data collection sheet

☐ use the total probability of 1 to calculate the probabilities of events

You are working at **Grade D** level.

☐ find an estimate of the mean from a grouped table of continuous data

☐ draw a frequency diagram for continuous data

☐ calculate the relative frequency of an event from experimental data

☐ interpret a scatter diagram

☐ use a line of best fit to predict values

☐ design and criticise questions for questionnaires

☐ describe the different methods of data collection

You are working at **Grade C** level.

☐ describe how to use the four stages of the data handling cycle to investigate a problem or a hypothesis

☐ draw a cumulative frequency diagram

☐ find the median, quartiles and interquartile range from a cumulative frequency diagram

☐ draw and interpret box plots

☐ draw a tree diagram to work out probabilities of combined events

You are working at **Grade B** level.

☐ draw histograms from frequency tables with unequal intervals

☐ calculate the numbers to be sampled for a stratified sample

☐ use AND/OR to work out the probabilities of combined events

You are working at **Grade A** level.

☐ find the median, quartiles and interquartile range from a histogram

☐ work out the probabilities of combined events when the probabilities change depending on previous outcomes (conditional probability)

You are working at **Grade A*** level.

Real-life problems

- In a GCSE examination, long multiplication and long division are usually given in the context of **real-life problems**.

- You will need to **identify** the problem as a **multiplication** or **division** and then work it out by your **preferred method**.

> **REMEMBER**
> Show your working clearly because even if you make a small arithmetical error, you will still get marks for method.

- In the GCSE non-calculator paper, questions often ask such things as 'How many coaches are needed?' and the calculation gives a remainder. Remember that an extra coach will be needed to carry the remaining passengers. You cannot have a 'bit of a coach'.

A café uses 950 eggs per day. The eggs come in trays of 24. How many trays of eggs will the café need?

This is a division problem. It is done using the traditional method.

```
      39
  24)950
     720
     230
     216
      14
```

The answer to the calculation 950 ÷ 24 is 39 remainder 14.

The café will need 40 trays and will have 10 eggs left over.

Work out 243 × 68.

Using the box method, split the two numbers into hundreds, tens and units, write them in a grid like the one below and multiply all the pairs.

×	200	40	3
60	12 000	2400	180
8	1 600	320	24

Add the separate answers to find the total.

12 000 + 2400 + 180 + 1600 + 320 + 24 = 16 524

So 243 × 68 = 16 524

Dividing by decimals

- To divide by a decimal, it is advisable to change the problem into one where you divide by an integer.

- The divisor and the dividend are multiplied by 10 or 100, etc.

Evaluate the following. **a** 34 ÷ 0.2 **b** 16.2 ÷ 0.45

a 34 ÷ 0.2 = 340 ÷ 2 = 170

b 16.2 ÷ 0.45 = 1620 ÷ 45, which then becomes a long division problem to which the answer is 36.

```
1620
 900    20 × 45
 720
 450    10 × 45
 270
 180    4 × 45
  90
  90    2 × 45
   0    36 × 45
```

Questions

Grade D

FM 1 **a** There are 945 students in a school. There are 27 students in each tutor group. How many tutor groups are there?

   b To raise money for charity, one tutor group decides that all 27 members will donate the cost of a KitKat. If a KitKat costs 42p, how much money do they raise?

Grade D

2 Work out each of the following.

   a 7.2 ÷ 0.4

   b 7.8 ÷ 0.6

   c 5.16 ÷ 0.43

Grade D

AU 3 How could you tell without working out the answer that these two calculations are wrong?

   a 53 × 65 = 3608 **b** 120 ÷ 250 = 0.84

Rounding and approximating

- To round a number to n significant figures, find the number that is closest with n digits and the rest of the places made up with zeros.

  > 432 is 400 to 1 sf and 430 to 2 sf.
  >
  > 0.08763 is 0.09 to 1 sf, 0.088 to 2 sf and 0.0876 to 3 sf.

- To **approximate** the answer to a calculation, round each number in the calculation to 1 significant figure, then work out the calculation.

  > $38.2 \times 9.6 \approx 40 \times 10 = 400$, so $38.2 \times 9.6 \approx 400$
  >
  > $48.3 \div 19.7 \approx 50 \div 20 = 2.5$, so $48.3 \div 19.7 \approx 2.5$

- The sign $\approx$ means 'is approximately equal to'.

REMEMBER
In your GCSE examination, you could be asked to round to 1, 2 or 3 significant figures. Final answers in exams should not be rounded to less than 3 significant figures unless you are instructed otherwise.

Multiplying and dividing by powers of 10

- When you **multiply** by a power of 10, the digits of the number move to the **left**.

- When you **divide** by a power of 10, the digits of the number move to the **right**.

- The number of **places the digits move** depends on the number of zeros or the power of 10.

REMEMBER
Although strictly speaking the digits move, it may be easier to think of it as the decimal point moving.

$45.9 \div 10^2 = 0.459$

10	1	$\frac{1}{10}$	$\frac{1}{100}$	$\frac{1}{1000}$	
4	5 . 9				$\div 10^2$
	. 4	5	9		

$45.9 \times 10^2 = 4590$

1000	100	10	1	$\frac{1}{10}$	
		4	5 . 9		$\times 10^2$
4	5	9	0 . 0		

Multiplying and dividing multiples of 10

- To **multiply** together two multiples of 10, multiply the non-zero digits and write the total of the zeros in both numbers at the end.

- To **divide** one multiple of 10 by another, divide the non-zero digits and write the difference in the zeros in both numbers at the end.

> $200 \times 4000 = 800\ 000$
>
> $500 \times 60 = 30\ 000$
>
> $8000 \div 20 = 400$
>
> $20\ 000 \div 40 = 500$

Questions

Grade C

1 a Round each of these numbers to 1 significant figure.

   i 3.8   **ii** 0.752   **iii** 58.7

b Round each of these numbers to 2 significant figures.

   i 56.8   **ii** 0.965   **iii** 88.9

Grade D

2 Find an approximate answer to each of the following.

   a 68.3×12.2 **b** $203.7 \div 38.1$

   c $\dfrac{78.3 + 19.6}{21.8 - 9.8}$ **d** $\dfrac{42.1 + 78.6}{24.7 - 19.3}$

Grade D

3 Write down the answer to each of the following.

   a 0.3×10^3 **b** 7.6×100 **c** $0.75 \div 10$

   d $34 \div 10^3$ **e** 3000×200 **f** $4000 \div 20$

Prime factors

- When a number is written as a product of **prime factors**, it is written as a multiplication consisting only of prime numbers.

$$30 = 2 \times 3 \times 5 \qquad 50 = 2 \times 5 \times 5 \text{ or } 2 \times 5^2$$

- To find the prime factors of a number, divide by prime numbers until the answer is a prime number.

Find the prime factors of 24.

Divide by prime numbers until the answer is a prime number.

So 24 = 2 × 2 × 2 × 3

```
2 ) 2 4
2 ) 1 2
2 )   6
      3
```

- Products of prime factors can be expressed in **index form**.

$$24 = 2 \times 2 \times 2 \times 3 = 2^3 \times 3 \qquad 76 = 2 \times 2 \times 19 = 2^2 \times 19$$

Lowest common multiple and highest common factor

- The **lowest common multiple (LCM)** of two numbers is the smallest number in the times tables of both of the numbers.

The LCM of 6 and 7 is 42.   The LCM of 8 and 20 is 40.

- The **highest common factor (HCF)** of two numbers is the biggest number that divides exactly into the two numbers.

The HCF of 24 and 18 is 6.   The HCF of 45 and 36 is 9.

- To find the LCM and HCF, write out the multiples and factors of each number.

Find the LCM and HCF of 16 and 20.

LCM: Write out the 16 and 20 times tables, continuing until there is a common multiple.

16: 16, 32, 48, 64, 80, 96, … 20: 20, 40, 60, 80, 100, … So the LCM of 16 and 20 is 80.

HCF: Write out the factors of 16 and 20 then pick out the biggest number that appears in both lists.

Factors of 16: {1, 2, 4, 8, 16} Factors of 20: {1, 2, 4, 5, 10, 20} So the HCF of 16 and 20 is 4.

Questions

Grade C

1 a What numbers are represented by these products of prime factors?

   i $2 \times 3 \times 5$ ii $2 \times 2 \times 3 \times 7$ iii $2 \times 5 \times 13$

   iv $2^2 \times 3^2$ v $2^3 \times 5$ vi $2^2 \times 3 \times 5^2$

  b Use the division method to find the prime factors of each of these numbers. Give your answers in index form where possible.

   i 20   ii 45   iii 64   iv 120

Grade C

2 a Find the LCM of the numbers in each pair.

   i 5 and 6   ii 3 and 7   iii 3 and 13

  (AU b) Describe a connection between the LCM and the original numbers in part **a**.

  c Find the LCM of each the numbers in each pair.

   i 6 and 9   ii 8 and 10   iii 15 and 25

  d Find the HCF of each of the numbers in each pair.

   i 12 and 30   ii 18 and 40   iii 15 and 50

   iv 16 and 80   v 24 and 60   vi 12 and 25

Fractions

One quantity as a fraction of another

- To write one quantity as a fraction of another, write the **first quantity** as the **numerator** and the **second quantity** as the **denominator**.

  What is £8 as a fraction of £20?

  Write as $\frac{8}{20}$ then cancel to $\frac{2}{5}$.

> **REMEMBER**
> Examination questions often ask for the fraction to be given in its simplest form. This means it has to be cancelled down.

Adding and subtracting fractions

- When **adding** and **subtracting** fractions, use **equivalent fractions** to make the denominators of the fractions the same.

  $\frac{3}{8} + \frac{1}{5} = \frac{3 \times 5}{8 \times 5} + \frac{1 \times 8}{5 \times 8} = \frac{15 + 8}{40} = \frac{23}{40}$

- When adding or subtracting mixed numbers, split the calculation into whole numbers and fractions.

  $4\frac{2}{5} - 1\frac{3}{4} = 4 - 1 + \frac{2}{5} - \frac{3}{4} = 3 + \frac{8}{20} - \frac{15}{20} = 3 + -\frac{7}{20} = 2\frac{13}{20}$

> **REMEMBER**
> Notice that you have to split one of the whole numbers if the answer to the fractional part is negative.

Multiplying and dividing fractions

- To **multiply** two fractions, multiply the numerators and multiply the denominators.

  $\frac{3}{4} \times \frac{3}{7} = \frac{3 \times 3}{4 \times 7} = \frac{9}{28}$

- When multiplying **mixed numbers**, change the mixed numbers into **top-heavy** fractions then multiply, as for ordinary fractions.

- **Cancel** any common factors in the top and bottom before multiplying.

- Convert the **final answer** back to a **mixed number** if necessary.

  $3\frac{1}{4} \times 1\frac{1}{5} = \frac{13}{4} \times \frac{\overset{3}{\cancel{6}}}{5} = \frac{13 \times 3}{2 \times 5} = \frac{39}{10} = 3\frac{9}{10}$

> **REMEMBER**
> On calculator papers you will be expected to use a calculator to do fraction problems, so learn how to use the fraction buttons on your calculator.

- To **divide** by a fraction, turn it upside down and multiply by it.

  $\frac{3}{4} \div \frac{7}{9} = \frac{3}{8} \times \frac{9}{7} = \frac{27}{56}$

- When dividing mixed numbers, change them into top-heavy fractions and then divide, as for ordinary fractions.

  When the second fraction has been turned upside down, cancel before multiplying.

  $1\frac{1}{4} \div 1\frac{7}{8} = \frac{5}{4} \div \frac{15}{8} = \frac{\overset{1}{\cancel{5}}}{\underset{1}{\cancel{4}}} \times \frac{\overset{2}{\cancel{8}}}{\underset{3}{\cancel{15}}} = \frac{2}{3}$

Questions

Grade D

1 a What fraction of 25 is 10?

  b In a class of 28 students, 21 are right-handed. What fraction is this?

  c What fraction is 20 minutes of one hour?

Grade D–C

2 Work out each of these.

  a i $\frac{1}{4} + \frac{3}{7}$ ii $\frac{5}{6} + \frac{4}{9}$ iii $3\frac{2}{3} + 2\frac{2}{5}$

  b i $\frac{3}{5} - \frac{1}{6}$ ii $\frac{8}{9} - \frac{2}{3}$ iii $2\frac{1}{4} - 1\frac{2}{3}$

  c i $\frac{3}{4} \times \frac{2}{9}$ ii $\frac{5}{8} \times \frac{4}{7}$ iii $1\frac{2}{5} \times 2\frac{3}{4}$

  d i $\frac{3}{5} \div \frac{6}{7}$ ii $\frac{5}{6} \div \frac{10}{21}$ iii $3\frac{3}{5} \div 2\frac{1}{4}$

Percentage

AQA 1/2   EDEXCEL 1/2/3   OCR 2/3

C

The percentage multiplier

- Using the **percentage multiplier** is the best way to solve percentage problems.
- The percentage multiplier is the **percentage** expressed as a **decimal**.

  72% gives a multiplier of 0.72.   20% gives a multiplier of 0.20 or 0.2.

- The multiplier for a percentage **increase or decrease** is the percentage multiplier **added to or subtracted from 1**.

  An 8% increase is a multiplier of 1.08 (1 + 0.08).

  A 5% decrease is a multiplier of 0.95 (1 – 0.05).

> **REMEMBER**
> Learn how to use multipliers as they make percentage calculations easier and more accurate.

C

Calculating a percentage increase or decrease

- To calculate the new value after a quantity is **increased** or **decreased** by a percentage, simply **multiply** the original quantity by the **percentage multiplier** for the increase or decrease.

  What is the new cost after a price of £56 is decreased by 15%?

  Work it out as 0.85 × 56 = £47.60.

> **REMEMBER**
> If the calculator shows a number such as 47.6 as the answer to a money problem, always put the extra zero into the answer, so you write this down as £47.60.

C

Expressing one quantity as a percentage of another

- To calculate one **quantity as a percentage of another**, divide the first quantity by the second. This will give a decimal, which can be converted to a percentage.

  A plant grows from 30 cm to 39 cm in a week. What is the percentage growth?

  The increase is 9 cm. 9 ÷ 30 = 0.3 and this is 30%.

> **REMEMBER**
> Always divide by the original quantity, otherwise you will not get any marks.

Questions

Grade D

1 a Write down the percentage multiplier for each of these.

   i 80%   ii 7%   iii 22%

   b Write down the multiplier for each percentage increase.

   i 5%   ii 12%   iii 3.2%

   c Write down the multiplier for each percentage decrease.

   i 8%   ii 15%   iii 4%

Grade D

2 a Increase £150 by 12%.

   b Decrease 72 kg by 8%.

Grade C

FM 3 a The average attendance at Barnsley football club in 2009 was 14 800.

   In 2010 it was 15 540.

   i By how much had the average attendance gone up?

   ii What is the percentage increase in attendance?

   b After her diet, Carol's weight had gone from 80 kg to 64 kg.

   What is the percentage decrease in her weight?

32 **Number**

Compound interest

- When money is **invested** in, for example, a savings account, it can earn a certain rate of **interest** each year.

- This interest is then added to the original amount and the **new total amount** then earns interest at the same rate in the following year.

- The process whereby interest is **accumulated** each year is called **compound interest**.

- The best way to calculate the compound interest is to use a **multiplier**.

> **REMEMBER**
> If the answer is a string of decimals, round off to something sensible, that is a whole number for discrete data or 2 decimal places for money.

> A bank pays 4% compound interest per year. Jack invests £3000. How much will he have after 5 years?
>
> The multiplier for an annual increase of 4% is 1.04
>
> The calculation is: $3000 \times 1.04^5 = £3649.96$

- Remember the formula as **total amount** $= P \times (1 + x)^n$, where P is the original amount invested, x is the interest rate expressed as a decimal and n is the number of years for which the money is invested.

- This type of problem can also be about increasing or decreasing populations, salaries, weights, etc.

> A Petri dish containing 20 000 bacteria is treated with a detergent which kills 12% of the bacteria each minute. How many bacteria will remain after 10 minutes?
>
> The multiplier for a decrease of 12% is 0.88
>
> The calculation is: $20\,000 \times 0.88^{10} = 5570$

Reverse percentage

- After an amount has been increased or decreased by a certain percentage, the original amount can be found from the new amount.

- There are two methods to do this: the **unitary method** and the **multiplier method**.

> **REMEMBER**
> The multiplier method is much simpler and is shown below.

> After a 22% increase, the population of a village is 1464. What was the population originally?
> The multiplier for a 22% increase is 1.22
> The calculation is: $1464 \div 1.22 = 1200$
> So the original population was 1200.

> **REMEMBER**
> Answers to percentage questions are usually sensible numbers. If you get a string of decimals such as 1141.92 for the population of a village, you have almost certainly made a mistake.

Questions

Grade C

FM 1 a £500 is invested in an account that pays 3.5% compound interest.

How much will be in the account after 6 years?

b A rabbit colony has a disease that reduces its population by 9% each year. If the original population was 2000, how many rabbits will there be after 4 years?

Grade B

FM 2 a In a sale, a TV is reduced to £322. This is an 8% reduction from the original price. What was the original price?

b The average attendance at Barnsley football club increased by 12% between 2009 and 2010.
In 2010 the average attendance was 15 680.
What was the average attendance in 2009?

Number

Ratio

AQA 1/2/3   EDEXCEL 1/2   OCR 1/3

Ratios

- A **ratio** is a way of comparing the sizes of two or more quantities.
- A colon (:) is used to show ratios.   3 : 4 and 6 : 20 are ratios.
- A quantity can be divided into **portions** that are in a **given ratio**.
- The process has three steps.
    - Step 1: **Add** the separate parts of the ratio
    - Step 2: **Divide** this number into the original quantity
    - Step 3: **Multiply** this answer by the original parts of the ratio

> Share £40 in the ratio 2 : 3.
> Add 2 and 3 to find the total number of parts: 5
> Divide 40 by 5 to find the value of each part: 8
> Multiply each term in the ratio by 8: 2 × 8 = 16, 3 × 8 = 24
> So £40 divided in the ratio 2 : 3 gives shares of £16 and £24.

REMEMBER

Always check that the two parts into which you have divided the quantity add up to the original amount.

- When one part of a ratio is known, it is possible to calculate other values. Use the given information to find a unit value and use the **unit value** to find the required information.

> When the cost of a meal was shared between two families in the ratio 3 : 5, the smaller share was £22.50. How much did the meal cost altogether?
> $\frac{3}{8}$ of the cost was £22.50, so $\frac{1}{8}$ was £7.50. The total cost was 8 × 7.50 = £60.

Speed, time and distance

REMEMBER

If you are using a calculator, make sure you convert minutes into decimals, for example 2 hours 15 minutes is 2.25 hours.

- **Speed**, **time** and **distance** are connected by the formula:
  distance = speed × time
- This formula can be rearranged to give: $\textbf{speed} = \dfrac{\textbf{distance}}{\textbf{time}}$ $\textbf{time} = \dfrac{\textbf{distance}}{\textbf{speed}}$
- Problems involving speed actually mean **average speed**, as maintaining a constant speed is not possible over a journey.

> A car travels at 40 mph for 2 hours. How far does it travel in total?
> distance = speed × time = 40 × 2 = 80 miles

- Use this diagram to remember the formulae that connect speed, time and distance.

Questions

Grade C

1 Divide the following amounts in the given ratios.
    a £500 in the ratio 1 : 4
    b 300 grams in the ratio 1 : 5
    c £400 in the ratio 3 : 5
    d 240 kg in the ratio 1 : 2

Grade C

2 a A box of crisps has two flavours, plain and beef, in the ratio 3 : 4. There are 42 packets of plain crisps. How many packets of beef crisps are there in the box?

   b The ratio of male teachers to female teachers in a school is 3 : 7. If there are 21 female teachers, how many teachers are there in total?

Grade D

FM 3 a A motorist travels a distance of 75 miles in 2 hours. What is his average speed?

   b A cyclist travels for $3\frac{1}{2}$ hours at an average speed of 15 km per hour.
   How far has she travelled?

Direct proportion problems

- When solving **direct proportion** problems, work out the **cost of one item**. This is called the **unitary method**.

> If eight cans of cola cost £3.60, how much do five cans cost?
>
> The cost of one can is 360 ÷ 8 = 45p, so five cans cost 5 × 0.45 = £2.25.

D

Best buys

- Many products are sold in different sizes at different prices.

- To find a **best buy**, work out how much of the item you get for a unit cost, such as how much you get per penny or how much per pound.

- Always divide the quantity by the cost.

> A 400 g jar of coffee costs £1.44. A kilogram jar of the same coffee costs £3.80. Which jar is better value?
>
> 400 ÷ 144 = 2.78 g/penny   1000 ÷ 380 = 2.63 g/penny
>
> Hence the 400 g jar is better value.

D

Density

- **Density** is the mass of a substance per unit volume.

- Density is usually expressed in grams per cubic centimetre (g/cm³) or kilograms per cubic metre (kg/m³).

- The relationship between density, mass and volume is:

$$\text{density} = \frac{\text{mass}}{\text{volume}}$$

- The following triangle can be used to help remember the formulae that connect mass, density and volume.

mass = density × volume
density = mass ÷ volume
volume = mass ÷ density

C

Questions

Grade D

FM 1 **a** 40 bricks weigh 50 kg. How much will 25 bricks weigh?

b How many bricks are there on a pallet weighing 200 kg, assuming the weight of the pallet is zero?

Grade D

FM 2 **a** A large tube of toothpaste weighs 250 g and costs £1.80. A travel-size tube contains 75 g and costs 52p. Which is better value?

b Which is the better mark: 62 out of 80 or 95 out of 120?

Grade C

FM 3 **a** Calculate the density of a brick with a mass of 2500 grams and a volume of 400 cm³. State the units in your answer.

b Calculate the weight, in kilograms, of a piece of metal 3000 cm³ in volume, if its density is 9 g/cm³.

c A piece of wood has an average density of 52 kg/m³. It weighs 15.6 kg.

Find its volume. Give your answer in cubic metres.

Powers and reciprocals

C

Square roots and cube roots

- The **square root** of a given number is a number which, when multiplied by itself, produces the given number.

  The square root of 16 is 4, since $4 \times 4 = 16$.

- Every positive number has two square roots, one positive and one negative.

  Because -4×-4 also equals 16, $\sqrt{16} = \pm 4$.

- A square root is represented by the **symbol** $\sqrt{}$.

  $\sqrt{25} = 5$

- Calculators have a '**square root**' button. $\boxed{\sqrt{}}$

- Taking a square root is the **inverse operation** to squaring.

- The **cube root** of a given number is a number which, when multiplied by itself twice, produces the given number.

  The cube root of 125 is 5, since $5 \times 5 \times 5 = 125$.

- A **cube root** is represented by the symbol $\sqrt[3]{}$.

  $\sqrt[3]{125} = 5$

- Many calculators have a '**cube root**' button. $\boxed{\sqrt[3]{}}$

C

Powers

- **Powers** are a convenient way of writing repetitive multiplications.

  $3 \times 3 \times 3 \times 3 \times 3 \times 3 = 3^6$, which is read as '3 to the power 6'.

- Powers are also called **indices** (singular **index**).

- The **power** '**2**' has a special name: '**squared**'.

- The **power** '**3**' has a special name: '**cubed**'.

Negative indices

B

- A negative index is a convenient way of writing the **reciprocal** of a number or term.

  $2^{-2} = \frac{1}{2^2} = \frac{1}{4}$ $\frac{1}{27} = \frac{1}{3^3} = 3^{-3}$

> **REMEMBER**
> A negative index does not mean the answer is negative. It means the reciprocal.

Questions

Grade D–C

1 a Write down the value of each of these.

   i $\sqrt{81}$ **ii** $\sqrt[3]{64}$ **iii** $\sqrt{25}$

 b Find the values of x in each of these expressions.

   i $x^2 = 4$ **ii** $x^3 = 1$ **iii** $x^3 = -8$

 c Use a calculator to find the value of each of these.

   i $\sqrt{576}$ **ii** $\sqrt[3]{1.331}$ **iii** $\sqrt{37.21}$

Grade D–C

2 a Write down the value of each of these cubes.

   i 3^3 **ii** 4^3 **iii** 10^3

 b Write these expressions, using power notation.

   i $4 \times 4 \times 4 \times 4 \times 4$

   ii $6 \times 6 \times 6 \times 6 \times 6 \times 6$

   iii $10 \times 10 \times 10 \times 10$

   iv $2 \times 2 \times 2 \times 2 \times 2 \times 2 \times 2$

 c Use a calculator to work out the values of the power terms in part **b**.

 d Continue the powers of 2 sequence for 10 terms.

   $2, 4, 8, 16, 32, \ldots, \ldots, \ldots, \ldots, \ldots$

Grade B

3 a Write down each of these in fraction form.

   i 4^{-3} **ii** 7^{-1} **iii** 3^{-2}

 b Write down each of these in negative index form.

   i $\frac{1}{8}$ **ii** $\frac{1}{3}$ **iii** $\frac{1}{x^n}$

Multiplying and dividing powers

- When we **multiply** together powers of the same number or variable, we **add** the indices.

$$2^2 \times 2^3 = 2^{(2+3)} = 2^5 = 32 \qquad x^2 \times x^6 = x^{(2+6)} = x^8$$

- When we **divide** two powers of the same number or variable, we **subtract** the indices.

$$3^5 \div 3^2 = 3^{(5-2)} = 3^3 = 27 \qquad x^9 \div x^6 = x^{(9-6)} = x^3$$

- When you are working with more complex expressions, the same rules apply. Treat each letter and number separately.

$$\frac{4a^3b^3 \times 2ab^2}{2a^2b} = \frac{4 \times 2 \times a^3 \times a \times b^3 \times b^2}{2a^2b} = \frac{8a^4b^5}{2a^2b} = 4a^2b^4$$

> **REMEMBER**
> Don't get confused with ordinary numbers when powers are involved. The rules above only apply to powers, so $2x^3 \times 3x^2$
> $= 2 \times 3 \times x^3 \times x^2 = 6x^5$
> not $5x^5$.

Power of a power

- When we **raise** a power term to a further power, we **multiply** the powers.

$$(a^2)^3 = a^6 \qquad (2x^3)^4 = 2^4 \times (x^3)^4 = 16x^{12}$$

Indices of the form $\frac{1}{n}$ and $\frac{a}{b}$

- An index of the form $\frac{1}{n}$ is the **nth root** of that number.

$$49^{\frac{1}{2}} = \sqrt{49} = 7 \qquad 8^{\frac{1}{3}} = \sqrt[3]{8} = 2 \qquad 36^{-\frac{1}{2}} \; \frac{1}{\sqrt{36}} = \frac{1}{6}$$

> **REMEMBER**
> A negative power does not mean the answer will be negative. It means take the reciprocal.

- An index of the form $\frac{a}{b}$ is the ath power of the bth root. $32^{\frac{4}{5}} = (32^{\frac{1}{5}})^4 = 2^4 = 16$

- If the power is **negative**, take the **reciprocal** of the answer.

- Solve these problems in three steps.

  – Step 1: Take the root of the base number given by the denominator of the fraction

  – Step 2: Raise the result to the power given by the numerator of the fraction

  – Step 3: Take the reciprocal (divide into 1) of the answer, which is what a negative power tells you to do

$$125^{-\frac{2}{3}} = (125^{\frac{1}{3}})^{-2} = (5)^{-2} = \frac{1}{25}$$

Questions

Grade C–B

1 a Write each of the following as a single power.

  i $2^3 \times 2^4$ **ii** $2^4 \times 2^5$ **iii** $x^6 \times x^3$

  iv $3^5 \div 3^2$ **v** $3^8 \div 3^4$ **vi** $x^7 \div x^3$

  b Simplify these expressions.

  i $\dfrac{2a^2bc^3 \times 4ab^2c}{2abc}$ **ii** $\dfrac{6xyz \times 2x^2y^3z^2 \times 3xz}{9x^2yz^3}$

Grade B–A

2 Simplify these expressions.

  a $(x^3)^5$ **b** $(3a^2)^3$ **c** $(2xy^2)^2$

Grade A

3 Evaluate the following.

  a $25^{\frac{1}{2}}$ **b** $64^{\frac{1}{3}}$ **c** $125^{-\frac{1}{3}}$

Grade A

4 Evaluate the following.

  a $64^{-\frac{2}{3}}$ **b** $27^{-\frac{2}{3}}$ **c** $100^{-\frac{5}{2}}$

B

Standard form

- **Standard form** (also called standard index form) is a way of writing very large or very small numbers, using powers of 10.

- Every standard form number is of the form $a \times 10^n$, where $1 \leq a < 10$ and n is a positive or negative whole number.

  $45\,000 = 4.5 \times 10^4$ $0.000\,089 = 8.9 \times 10^{-5}$

- Standard form numbers can be combined, using the **rules of powers**.

  Work out $3 \times 10^2 \times 5 \times 10^3$.
  $3 \times 10^2 \times 5 \times 10^3 = 3 \times 5 \times 10^2 \times 10^3 = 15 \times 10^5 = 1.5 \times 10^6$

  Work out $(4.5 \times 10^3) \div (5 \times 10^7)$.
  $(4.5 \times 10^3) \div (5 \times 10^7) = 4.5 \div 5 \times 10^3 \div 10^7 = 0.9 \times 10^{-4} = 9 \times 10^{-5}$

> **REMEMBER**
> Count how many places you have to move the decimal point to get the numerical value of the power. Moving right is negative, left is positive.

> **REMEMBER**
> Make sure your final answer is in standard form. You will lose a mark if it isn't.

C

Rational numbers

- A **rational number** is any number that can be expressed as a **fraction**.

- Some fractions result in **terminating decimals** and some fractions result in **recurring decimals**.

  $\frac{1}{16} = 0.0625$ which is a terminating decimal.

  $\frac{1}{3} = 0.3333...$ which is a recurring decimal.

- Recurrence is shown by a dot or dots over the recurring digit or digits.

  $0.3333...$ becomes $0.\dot{3}$ $0.277\,777...$ becomes $0.2\dot{7}$
  $0.518\,518\,518...$ becomes $0.\dot{5}1\dot{8}$

> **REMEMBER**
> The only fractions that give terminating decimals are those with a denominator that is a power of 2, 5 or 10, or a combination of these.
> $\frac{3}{64} = 0.046\,875$
> $\frac{7}{40} = 0.175$
> $\frac{19}{1000} = 0.019$

C

Finding reciprocals

- The **reciprocal** of a number is the result of **dividing** the number **into 1**.

- The reciprocal of a **fraction** is simply the fraction turned **upside down**.

The reciprocal of 5 is $\frac{1}{5}$ or $1 \div 5 = 0.2$.

The reciprocal of $\frac{4}{7}$ is $\frac{7}{4} = 1\frac{3}{4}$.

Questions

Grade B–A

1 a Write these numbers in standard form.
   i 560 000   **ii** 0.000 007   **iii** 3 million

b Write each of these as an ordinary number.
   i 6.4×10^6 **ii** 8.3×10^{-4} **iii** 9×10^8

c Work out the following. Give your answers in standard form.
   i $5.2 \times 10^4 \times 3 \times 10^2$
   ii $(3.6 \times 10^6) \div (9 \times 10^2)$
   iii $1.8 \times 10^2 \times 5 \times 10^4$
   iv $(2.4 \times 10^4) \div (3 \times 10^8)$

Grade C

2 Write each of the following using recurrence notation.
   a 0.363 636...   **b** 0.615 615 615...
   c 0.366 6...   **d** $\frac{2}{3}$ **e** $\frac{1}{6}$ **f** $\frac{7}{9}$

Grade C

3 a Work out the reciprocal of each number. Give your answers as terminating or recurring decimals.
   i 4   **ii** 20   **iii** 9

b Write down the reciprocals of each fraction. Give your answers as mixed numbers.
   i $\frac{7}{8}$ **ii** $\frac{5}{9}$ **iii** $\frac{3}{13}$

Surds

Surds

- **Surds** are numerical expressions that contain square roots. $\sqrt{2}$ and $\sqrt{5}$ are surds.

- Answers given as surds are **exact** answers. Answers written as $\sqrt{5} \approx 2.236$ are approximations.

- There are four rules that must be followed when working with surds.

  $$\sqrt{2} \times \sqrt{6} = \sqrt{12} \qquad \sqrt{10} \div \sqrt{2} = \sqrt{5}$$
  $$3\sqrt{7} \times 2\sqrt{3} = 6\sqrt{21}$$
  $$8\sqrt{15} \div 2\sqrt{3} = 4\sqrt{5}$$

  - $\sqrt{a} \times \sqrt{b} = \sqrt{ab}$ - $C\sqrt{a} \times D\sqrt{b} = CD\sqrt{ab}$
  - $\sqrt{a} \div \sqrt{b} = \sqrt{\frac{a}{b}}$ - $C\sqrt{a} \div D\sqrt{b} = \frac{C}{D}\sqrt{\frac{a}{b}}$

- Expressions involving surds can usually be simplified.

  $$\sqrt{20} = \sqrt{4 \times 5} = 2\sqrt{5} \qquad \sqrt{20} + \sqrt{45} = 2\sqrt{5} + 3\sqrt{5} = 5\sqrt{5}$$
  $$(2 + \sqrt{3})(2 - \sqrt{12}) = 4 - 2\sqrt{12} + 2\sqrt{3} - \sqrt{36}$$
  $$= 4 - 4\sqrt{3} + 2\sqrt{3} - 6 = -2 - 2\sqrt{3}$$
  $$(4 + \sqrt{5})(4 - \sqrt{5}) = 16 - 4\sqrt{5} + 4\sqrt{5} - 5 = 11$$
  $$(2 + \sqrt{6})^2 = (2 + \sqrt{6})(2 + \sqrt{6}) = 4 + 2\sqrt{6} + 2\sqrt{6} + 6$$
  $$= 10 + 4\sqrt{6}$$

> **REMEMBER**
> Try to split numbers into a product that involves a square number.

> **REMEMBER**
> When you see a bracket squared, always write the bracket down twice.

Rationalising the denominator

- A denominator in surd form can be rationalised.

  Rationalise the denominator of $\frac{2}{\sqrt{5}}$.

  Multiply the top and the bottom by $\sqrt{5}$: $\frac{2 \times \sqrt{5}}{\sqrt{5} \times \sqrt{5}} = \frac{2\sqrt{5}}{5}$

  Rationalise the denominator of $\frac{4\sqrt{2}}{\sqrt{6}}$.

  Multiply the top and the bottom by $\sqrt{6}$: $\frac{4\sqrt{2} \times \sqrt{6}}{\sqrt{6} \times \sqrt{6}} = \frac{4\sqrt{12}}{6} = \frac{4 \times 2\sqrt{3}}{6} = \frac{4\sqrt{3}}{3}$

> **REMEMBER**
> Multiply the top and bottom by the same surd that is in the denominator.

Solving problems with surds

- Surds can be used to solve problems.

  Show that this triangle is right-angled.

  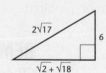

  If the triangle is right-angled then it will obey Pythagoras' theorem. (See pages 45–8.)

  $$(\sqrt{2} + \sqrt{18})^2 + 6^2 = (\sqrt{2} + \sqrt{18})(\sqrt{2} + \sqrt{18}) + 36$$
  $$= 2 + \sqrt{36} + \sqrt{36} + 18 + 36 = 68$$
  $$\sqrt{68} = \sqrt{(4 \times 17)} = 2\sqrt{17}. \text{ Hence it is right-angled.}$$

Questions

Grade A–A*

1 Simplify each of the following. Leave your answers in surd form.

a $\sqrt{12} \div \sqrt{3}$ b $2\sqrt{18} \times 3\sqrt{2}$ c $\sqrt{75}$

d $\sqrt{32} + \sqrt{50}$ e $\sqrt{3}(2 + \sqrt{3})$

f $(2 + \sqrt{5})(2 - \sqrt{5})$ g $(3 - \sqrt{5})^2$

Grade A*

2 Work out the area of a rectangle with length $2 + \sqrt{3}$ and width $5 - \sqrt{3}$. Give your answer in surd form.

Grade A*

3 Rationalise the denominators and simplify if possible.

a $\frac{3}{\sqrt{12}}$ b $\frac{3\sqrt{2}}{\sqrt{8}}$ c $\frac{3 - \sqrt{5}}{\sqrt{5}}$

Grade A*

AU 4 If $\sqrt{\frac{a}{b}} \div \sqrt{\frac{c}{d}} = N$, where N is a positive whole number, what relationship must exist between a, b, c and d?

Variation

Direct variation

A

- **Direct variation** means the same as **direct proportion**.

- There is direct variation between two variables when one variable is a simple multiple of the other.

- When two variables are proportional to each other, you can set up an equation with a constant of proportionality.

> Pay is directly proportional to time worked.
> Pay $\propto$ Time $\Rightarrow$ Pay = k × Time,
> where k is the constant of proportionality.

- When a power or root of one variable is directly proportional to another variable, the process is the same as for a linear direct variation.

> y is directly proportional to the square of x.
> $y \propto x^2 \Rightarrow y = kx^2$, where k is the constant of proportionality.

- In an examination question, some numerical information will be given to enable the constant of proportionality to be found.

> y is directly proportional to the square root of x.
> When $y = 6, x = 4$.
> $y \propto \sqrt{x} \Rightarrow y = k\sqrt{x}$. $6 = k\sqrt{4} \Rightarrow 6 = 2k \Rightarrow k = 3$.

- A question will usually give the value of one variable and ask for the value of the other variable.

> y is directly proportional to the cube of x.
> If $y = 20$ when $x = 2$, what is the value of y when $x = 4$?
> $y \propto x^3 \Rightarrow y = kx^3$. $20 = k × 2^3 \Rightarrow 20 = 8k \Rightarrow k = 2.5$.
> When $x = 4$, $y = 2.5 × 4^3 = 2.5 × 64 = 160$.

Inverse variation

A

- There is inverse variation between two variables when one variable decreases as the other increases such that the product of the two variables is constant.

- This is represented mathematically by the **reciprocal**.

> y is inversely proportional to x. $y \propto \frac{1}{x} \Rightarrow y = \frac{k}{x}$.

- Inverse variation problems involving powers, roots and given numerical information are solved using the same method as for direct variation problems.

> y is inversely proportional to the cube root of x.
> If $y = 20$ when $x = 8$, what is the value of y when $x = 64$?
> $y \propto \frac{1}{\sqrt[3]{x}} \Rightarrow y = \frac{k}{\sqrt[3]{x}} \Rightarrow 20 = \frac{k}{2} \Rightarrow k = 40$.
> When $x = 64$, $y = \frac{40}{\sqrt[3]{64}} = \frac{40}{4} = 10$.

> **REMEMBER**
> In a GCSE examination, writing down the proportionality equation gets a mark.

Questions

Grade A

1 y is directly proportional to x^2. When $y = 5$, $x = 10$.

  a Write down the equation of proportionality.

  b Find the constant of proportionality.

  c Find the value of y when $x = 5$.

  d Find the value of x when $y = 20$.

Grade A

2 y is inversely proportional to $\sqrt{x}$. When $y = 6$, $x = 4$.

  a Write down the equation of proportionality.

  b Find the constant of proportionality.

  c Find the value of y when $x = 9$.

  d Find the value of x when $y = 12$.

Limits of accuracy

- All recorded measurements will have been rounded off to some degree of accuracy.
- This defines the possible true values before the rounding took place, and hence the **limits of accuracy**.
- The range of values between the limits of accuracy is called the **rounding error**.

> The length of a piece of paper is 24.7 cm accurate to one decimal place.
> What are the limits of accuracy? $24.65 \leqslant$ length < 24.75
>
> The length of a motorway is 220 km to three significant figures.
> What are the limits of accuracy? $219.5 \leqslant$ length < 220.5

REMEMBER

The limits of accuracy are always given to one more degree of accuracy than the rounded value.

- The examples above all relate to **continuous** data. If the data is **discrete**, the limits of accuracy are different.

> The number of sweets in a jar is 100 to the nearest 10. What is the smallest and the greatest number of sweets that can be in the jar?
>
> The least is 95 and the most is 104.

Calculating with limits of accuracy

- When we combine two or more linear values, the errors in the linear measures will be compounded producing a larger error in the calculated value.
- It is important to combine the values together in the correct way. The table below shows the combinations to give the maximum and minimum values for the four rules of arithmetic.

Operation	Maximum	Minimum
Addition $(a + b)$	$a_{max} + b_{max}$	$a_{min} + b_{min}$
Subtraction $(a - b)$	$a_{max} - b_{min}$	$a_{min} - b_{max}$
Multiplication $(a \times b)$	$a_{max} \times b_{max}$	$a_{min} \times b_{min}$
Division $(a \div b)$	$a_{max} \div b_{min}$	$a_{min} \div b_{max}$

REMEMBER

Answers often come out with several decimal places. Never round off to less than three significant figures.

> The distance from Penistone to Glossop is 16.8 miles to the nearest tenth of a mile. The time for the journey calculated on a GPS unit is 25 minutes to the nearest minute. Between what limits is the expected average speed of the journey? Speed = distance ÷ time.
>
> maximum speed = maximum distance ÷ minimum time (hours) = $16.85 \div \frac{24.5}{60} = 41.27$ mph
>
> minimum speed = minimum distance ÷ maximum time (hours) = $16.75 \div \frac{25.5}{60} = 39.41$ mph
>
> The limits are $39.41 \leqslant$ speed < 41.27.

Questions

Grade C–A

1 Write down the lower and upper limits of each of these values, rounded to the accuracy stated.

  a 7 m (1 significant figures)

  b 34 kg (nearest kg)

  c 320 cm (2 significant figures)

Grade A

2 **a** A rectangle is measured as 8 cm by 12 cm, measured to the nearest cm. What are the limits of the area of the rectangle?

  (FM **b**) A journey of 32 miles, measured to the nearest mile, takes 60 minutes to the nearest 10 minutes. What are the limits of the average speed of the journey?

Number grade booster

I can...

☐ work out one quantity as a fraction of another
☐ solve problems using negative numbers
☐ multiply and divide by powers of 10
☐ multiply together numbers that are multiples of powers of 10
☐ round numbers to one significant figure
☐ estimate the answer to a calculation
☐ order lists of numbers containing decimals, fractions and percentages
☐ multiply and divide fractions
☐ calculate with speed, distance and time
☐ compare prices to find 'best buys'
☐ find the new value after a percentage increase or decrease
☐ find one quantity as a percentage of another

You are working at **Grade D** level.

☐ work out a reciprocal
☐ recognise and work out terminating and recurring decimals
☐ write a number as a product of prime factors
☐ find the HCF and LCM of pairs of numbers
☐ use the index laws to simplify calculations and expressions
☐ multiply and divide with negative numbers
☐ multiply and divide with mixed numbers
☐ round numbers to given numbers of significant figures
☐ find a percentage increase
☐ work out compound interest problems
☐ solve problems using ratio

You are working at **Grade C** level.

☐ work out the square roots of decimal numbers
☐ estimate answers using square roots of decimal numbers
☐ work out reverse percentage problems
☐ solve problems involving density
☐ write and calculate with numbers in standard form
☐ find limits of numbers given to the nearest unit

You are working a **Grade B** level.

☐ solve complex problems involving percentage increases and decreases
☐ use the rules of indices for fractional and negative indices
☐ convert recurring decimals to fractions
☐ simplify surds
☐ find limits of numbers given to various accuracies

You are working at **Grade A** level.

☐ solve problems using surds
☐ solve problems using combinations of numbers rounded to various limits

You are working at **Grade A*** level.

Circles and area

AQA 3   EDEXCEL 3   OCR 3

Circumference of a circle

- The **circumference** of a circle is the distance around the circle (the **perimeter**).

- The circumference of a circle is given by the formula:

  $C = \pi d$ or $C = 2\pi r$

  where d is the **diameter** and r is the **radius**.

REMEMBER
You can learn just one formula because you can always find the radius from the diameter, or vice versa, as: $d = 2r$

Calculate the circumference of the circle shown.
Give your answer to 3 significant figures.

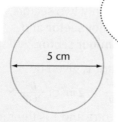

5 cm

$C = \pi d$
  $= \pi \times 5$ cm $= 15.7$ cm (to 3 sf)

- The value of π is a **decimal** that goes on forever but it is 3.142 to three decimal places.

- All calculators should have a π **button** which will give the value as 3.141592654.

- The formula you use depends on whether you are given the radius or the diameter.

- On the non-calculator paper, you may be asked to give an answer in terms of π. In this case leave your answer as, for example, 6π.

Area of a circle

REMEMBER
Unless you are asked to give your answer in terms of π, use a calculator to work out circumferences and areas of circles and give answers correct to at least one decimal place.

- The area of a circle is the **space inside** the circle.

- The area of a circle is given by the formula: $A = \pi r^2$ where r is the radius.

- You must use the radius when calculating the area.

8 cm   4 cm

Area of a trapezium

- To work out the area of a trapezium, multiply half the sum of the parallel sides by the distance between them using the formula:

  area $= \frac{1}{2} \times (a + b) \times h$

Find the area of this trapezium.
$A = \frac{1}{2} \times (8 + 12) \times 6$
  $= \frac{1}{2} \times 20 \times 6$
  $= 60$ m²

8 m
6 m
12 m

Questions

Grade D

1 **a** Calculate the circumference of a circle with a diameter of 12 cm.

   Give your answer to 1 decimal place.

   b Calculate the circumference of a circle with a radius of 4 cm.

   Leave your answer in terms of π.

Grade D

2 **a** Calculate the area of a circle with a radius of 15 cm. Give your answer to 1 decimal place.

b Calculate the area of a circle with a radius of 3 cm.

   Leave your answer in terms of π.

Grade D

3 Find the area of each of these trapezia.

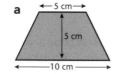

a 5 cm / 5 cm / 10 cm

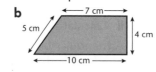

b 7 cm / 5 cm / 4 cm / 10 cm

Geometry

43

Sectors and prisms

Sectors

A

- A **sector** is part of a circle, bounded by two radii of the circle and one of the **arcs** formed by the intersection of these radii with the circumference.

- When a circle is divided into only two sectors, the larger one is called the **major sector** and the smaller one is called the **minor sector**.

- The length of the arc of a sector is given by the formula: **arc length** $= \frac{\theta°}{360°} \times 2\pi r$ or $\frac{\theta°}{360°} \times \pi d$ where θ is the angle at the centre and r is the radius.

- The area of a sector is given by the formula: **sector area** $= \frac{\theta°}{360°} \times \pi r^2$

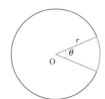

REMEMBER

On the non-calculator paper, you may be asked to give an answer in terms of π. In this case, the sector angle will be a factor of 360 such as 18, 20, or 30.

> Find the arc length and the area of the sector in the diagram.
>
> The sector angle is 35° and the radius is 6 cm. Therefore,
>
> arc length $= \frac{35°}{360°} \times \pi \times 2 \times 6 = 3.67$ cm
>
> sector area $= \frac{35°}{360°} \times \pi \times 6^2 = 11.0$ cm^2

Prisms

C

- A **prism** is a 3-D shape that has the same cross-section all the way through it.

- The **surface area** of a prism is the area covered by its net.

- The **volume** of a prism is found by multiplying its cross-sectional area by the length of the prism:
volume of a prism = **area of cross-section** × **length** or $V = Al$

> Calculate the surface area and the volume of this triangular prism.
>
> The surface area is made up of the area of the three rectangles and the area of the two isosceles triangles.
>
> Area of the three rectangles $= 10 \times 5 + 10 \times 5 + 10 \times 6$
> $= 50 + 50 + 60 = 160$ cm^2
>
> Area of one triangle $= \frac{6}{2} \times 4 = 12$ cm^2, so area of two triangles $= 24$ cm^2
>
> Therefore, the total surface area is 184 cm^2.
>
> $V = Al$. Area of the cross-section = area of one triangle = 12 cm^2.
>
> So, $V = 12 \times 10 = 120$ cm^3

Questions

Grade C

1 **a** A cuboid has dimensions 4 cm by 6 cm by 10 cm.

   i Calculate the area of its cross-section.

   ii Calculate its volume.

  b A triangular prism has a cross-sectional area of 3.5 m^2 and a length of 12 m. Calculate its volume.

Grade A

2 **a** Calculate the arc length and area of this sector.

  b A sector of a circle, radius 6 cm, has an angle of 40°.

   i Calculate the perimeter of the sector.

   ii Calculate the area of the sector.

  Give your answers in terms of π.

Cylinders and pyramids

AQA 3   EDEXCEL 2   OCR 3

Volume of a cylinder

- A **cylinder** is a **prism** with a **circular cross-section**.

- The formula for the **volume of a cylinder** is:

  $V = \pi r^2 h$

  where r is the radius and h is the height or length of the cylinder.

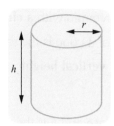

> **REMEMBER**
> You should learn the formula $V = \pi r^2 h$. It is not given on the formula sheet.

Surface area of a cylinder

- The **surface area** of a cylinder is made up of its **curved surface area** plus the area of its two **circular ends**.

- The curved surface opens out to a rectangle with length equal to the circumference of the circular end.

- The two circular ends each have an area of πr^2.

- The formula for the total surface area is:

  total surface area $= 2\pi rh + 2\pi r^2$ or $\pi dh + 2\pi r^2$

> What is the total surface area of a cylinder with a radius of 5 cm and a height of 12 cm?
> Total surface area $= 2 \times \pi \times 5 \times 12 + 2 \times \pi \times 5^2 = 534.1$ cm^2

Volume of a pyramid

- A **pyramid** is a 3-D shape with a base from which triangular faces rise to a common **vertex**.

- The base can be any **polygon**, but is usually a **triangle**, **square** or **rectangle**.

- The **volume** of a pyramid is given by the formula:

  $V = \frac{1}{3}Ah$

  where A is the base area and h is the vertical height.

- A triangular-based pyramid is called a **tetrahedron**.

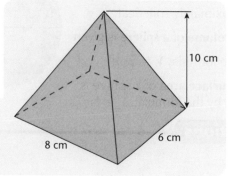

> Calculate the volume of the pyramid on the right.
> Base area $= 6 \times 8 = 48$ cm^2
> Volume $= \frac{1}{3} \times 48 \times 10 = 160$ cm^3

Questions

Grade C

1 Calculate the volume of a cylinder with a radius of 4 cm and a height of 10 cm.

   Give your answer in terms of π.

Grade A

2 Calculate the surface area of the cylinder in question **1**.

   Give your answer in terms of π.

Grade A

3 Calculate the volume of a pyramid with a square base of side 11 cm and a vertical height of 15 cm.

Grade A*

PS 4 The base of a pyramid is a square of side 10 cm. The total surface area of the pyramid is 300 cm^2. Calculate its volume.

Geometry

Cones and spheres

Cones

- A cone can be treated as a pyramid with a **circular base**.
- The formula for the **volume** of a cone is the same as that for a pyramid:

  Volume $= \frac{1}{3} \times$ base area $\times$ vertical height

  $$V = \frac{1}{3}\pi r^2 h$$

  where r is the radius of the base and h is the vertical height of the cone.

- The **curved surface area** of a cone is given by:

  $$S = \pi r l$$

  where l is the **slant height** of the cone.

- The **total surface area** of a cone consists of the curved surface area plus the area of its circular base.

- The **total surface area** is given by:

  $$A = \pi r l + \pi r^2$$

> **REMEMBER**
> The formula for the volume and curved surface area of a cone are given on the formula sheet that is included with the examination. However, it is much better to learn them.

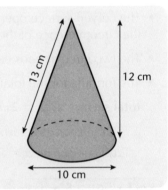

Calculate the volume and the total surface area of the cone shown.

Volume $\frac{1}{3} \times \pi \times 5^2 \times 12 = 314.2 \text{ cm}^3$

Total surface area $= \pi \times 5 \times 13 + \pi \times 5^2 = 282.7 \text{ cm}^2$

Spheres

- A sphere is a **ball** shape.
- The Earth and other planets are approximately spherical.
- The **volume** of a **sphere** is given by the formula: $V = \frac{4}{3}\pi r^3$
- The **surface area** of a sphere is given by the formula: $A = 4\pi r^2$

Calculate the volume and the surface area of a sphere with a radius of 9 cm.

Volume $= \frac{4}{3} \times \pi \times 9^3 = 3053.6 \text{ cm}^3$

Surface area $= 4 \times \pi \times 9^2 = 1017.9 \text{ cm}^2$

Questions

Grade A

1 For a cone with a radius of 6 cm, a vertical height of 8 cm and a slant height of 10 cm, calculate

   i its volume and   **ii** its curved surface area.

Grade A

2 For a sphere with a radius of 3 cm, calculate

   i its volume and   **ii** its surface area.

   Give your answers in terms of π.

Grade A

3 For a hemisphere with a radius of 5 cm, calculate

   i its volume and   **ii** its **total** surface area.

Grade A*

PS 4 A cone of height 10 cm is cut in two so that the small cone and frustum both have a height of 5 cm.

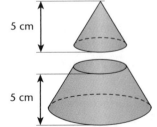

Explain why the ratio of the volume of the small cone to the volume of the frustum is 1 : 7.

Pythagoras' theorem

AQA 3   EDEXCEL 3   OCR 1

Pythagoras' theorem

- **Pythagoras' theorem** connects the sides of a **right-angled triangle**.

- Pythagoras' theorem states that:

  'In any right-angled triangle, the square of the hypotenuse is equal to the sum of the squares of the other two sides.'

- The **hypotenuse** is the **longest** side of the triangle, which is always **opposite** the right angle.

- Pythagoras' theorem is usually expressed as a formula: $c^2 = a^2 + b^2$

- The formula can be rearranged to find one of the other sides, for example: $a^2 = c^2 - b^2$

> **REMEMBER**
> To find the actual value of a side, don't forget to take the square root.

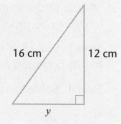

Finding lengths of sides

- If you **know the lengths of two sides** of a right-angled triangle, you can always use Pythagoras' theorem to **find the length of the third side**.

Find the length of the hypotenuse of this triangle.

Using Pythagoras' theorem

$x^2 = 9^2 + 6.2^2$
$= 81 + 38.44$
$= 119.44$
$x = \sqrt{119.44}$
$= 10.9$ cm

x, 6.2 cm, 9 cm

Find the length of the side marked *y* in this triangle.

Using Pythagoras' theorem

$y^2 = 16^2 - 12^2$
$= 256 - 144$
$= 112$
$y = \sqrt{112}$
$= 10.6$ cm

16 cm, 12 cm, *y*

Real-life problems

- Pythagoras' theorem can be used to solve practical problems.

A ladder of length 5 m is placed with the foot 2.2 m from the base of a wall. How high up the wall does the ladder reach?

$x^2 = 5^2 - 2.2^2$
$= 25 - 4.84$
$= 20.16$
$x = \sqrt{20.16} = 4.49$ m

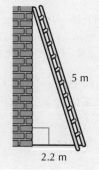

5 m

2.2 m

Questions

Grade C

1 Explain why a triangle with sides of 3 cm, 4 cm and 5 cm is right-angled.

Grade C

2 **a** Work out the length of the diagonal of a rectangle with sides 10 cm and 5 cm.

   b A ladder is placed 1.5 m from the base of a wall and reaches 3.6 m up the wall. How long is the ladder?

Grade B

(PS 3) Work out the length BC in this triangle.

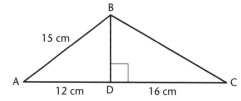

B, 15 cm, A, 12 cm, D, 16 cm, C

Pythagoras' theorem and isosceles triangles

- Every **isosceles triangle** has a line of symmetry that divides it into **two congruent** right-angled triangles.

Find the area of this triangle.

Split the triangle into two right-angled triangles with sides of 3.5 cm and 9 cm.

Use Pythagoras' theorem to find the height (h): $h = \sqrt{(9^2 - 3.5^2)} = 8.29$ cm

Area $= \frac{1}{2} \times 7 \times 8.29 = 29.0$ cm^2

Pythagoras' theorem in three dimensions

- 3-D problems in GCSE examinations can be solved using right-angled triangles and trigonometry (see page 47).

- Solve these problems in four steps.

  - Step 1: Identify the right-angled triangle that includes the required information

  - Step 2: Redraw and label this triangle with the given lengths and the length to be found

  - Step 3: Use your diagram to decide whether it is the hypotenuse or one of the shorter sides which has to be found

  - Step 4: Solve the problem and round to a suitable degree of accuracy

> **REMEMBER**
> The length of a diagonal in any cuboid with sides a, b and c is given by:
> $\sqrt{(a^2 + b^2 + c^2)}$

Find the distance BH in this cuboid.

First identify a right-angled triangle that contains the side BH and draw it.

This gives triangle BEH, which contains two lengths that you do not know, EB and BH. Let EB $= x$ and BH $= y$.

Next identify a right-angled triangle that contains the side EB and draw it.

This gives triangle EAB. You can now find EB.

By Pythagoras' theorem

$x^2 = 15^2 + 25^2 = 850$ (there is no need to find x)

Use triangle BEH to find BH.

By Pythagoras' theorem

$y^2 = 10^2 + 850 = 950$

So $y = BH = \sqrt{950} = 30.8$ cm

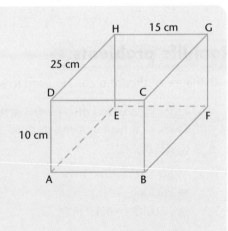

Questions

Grade B

1 Find the area of an isosceles triangle with sides of 10 cm, 10 cm and 8 cm.

Grade A

2 Find the length of a diagonal in a cuboid with sides of 3 cm, 4 cm and 5 cm.

Geometry

Using Pythagoras and trigonometry

Trigonometry

- **Trigonometry** is concerned with the calculation of sides and angles in triangles.
- There are three trigonometric ratios: **sine**, **cosine** and **tangent**. These are abbreviated to sin, cos and tan.
- The sine, cosine and tangent ratios are defined in terms of the sides of a right-angled triangle and an angle, θ:
- Calculators have **sine**, **cosine** and **tangent buttons**.
- To calculate angles, use the **inverse functions** on your calculator, usually marked $\sin^{-1}$, etc.

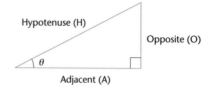

$\sin 45° = 0.707 \quad \cos^{-1} 0.4 = 66.4° \quad \tan^{-1} 0.5 = 26.6°$

Sine

- There are three possible rearrangements of the definition of sine:

$$\sin\theta = \frac{\text{opp}}{\text{hyp}} \qquad \text{opp} = \text{hyp} \times \sin\theta \qquad \text{hyp} = \frac{\text{opp}}{\sin\theta}$$

Use the sine ratio to find the value of x in this triangle.

x is opposite the known angle, so use opp $= \text{hyp} \times \sin\theta$

$x = 6 \times \sin 35° = 3.44$ cm

> **REMEMBER**
> All calculators work differently. Make sure you can use yours to get the answer shown.

Cosine

- There are three possible rearrangements of the definition of cosine:

$$\cos\theta = \frac{\text{adj}}{\text{hyp}} \qquad \text{adj} = \text{hyp} \times \cos\theta \qquad \text{hyp} = \frac{\text{adj}}{\cos\theta}$$

Use the cosine ratio to find the value of x in this triangle.

x is the hypotenuse, so use hyp $= \dfrac{\text{adj}}{\cos\theta}$

$x = 8 \div \cos 42° = 10.8$ cm

> **REMEMBER**
> Make sure your calculator is set to degrees. You will lose marks for using other angle measures.

Tangent

- There are three possible rearrangements of the definition of tangent:

$$\tan\theta = \frac{\text{opp}}{\text{adj}} \qquad \text{opp} = \text{adj} \times \tan\theta \qquad \text{adj} = \frac{\text{opp}}{\tan\theta}$$

Use the tangent ratio to find the angle x in this triangle.

x is the angle, so use $\tan\theta = \dfrac{\text{opp}}{\text{adj}}$

$\tan x = 7 \div 10 = 0.7$, so $x = \tan^{-1} 0.7 = 35.0°$

> **REMEMBER**
> The definitions of sine, cosine and tangent are not given in the formula sheets, so you need to learn them. A lot of people use SOHCAHTOA as a mnemonic.

Questions

Grade B

1 a Use your calculator to write down, to 3 significant figures, the value of:

   i $\sin 31°$ **ii** $\cos 55°$ **iii** $\tan 10°$

b Use your calculator to write down, to 1 decimal place, the value of:

   i $\sin^{-1} 0.6$ **ii** $\cos^{-1} 0.7$ **iii** $\tan^{-1} 2$

Grade B

2 Use the appropriate trigonometric ratio to find the values marked x in the triangles below.

Which ratio to use

- The difficulty with any trigonometric problem is knowing which ratio to use to solve it.

- If the problem involves **two sides and no angles**, use **Pythagoras' theorem** to find the missing side.

- If the problem has **an angle** in it, then you will need to use **trigonometry**.

- There are four steps when solving a trigonometry problem.

  - Step 1: Identify what information is given and what information needs to be found

  - Step 2: Decide which ratio to use

  - Step 3: Set up the appropriate calculation

  - Step 4: Work out the answer and round to an appropriate degree of accuracy

> **REMEMBER**
> Know your calculator. On some calculators it is better to type in $7 \times \cos 30$ and on others $\cos 30 \times 7$.

Find the length of the side marked x in this triangle.

Step 1: Identify what information is given and what needs to be found. Namely, x is adjacent to the angle and 7 cm is the hypotenuse.

Step 2: Decide which ratio to use. Only one ratio uses adjacent and hypotenuse: **cosine.**

Step 3: Set up the calculation. Remember adj = hyp $\times \cos \theta$, so the calculation is $x = 7 \times \cos 30°$.

Step 4: Work it out. $x = 6.06$ cm.

Solving problems

- Trigonometric problems in GCSE are usually set in real-life situations.

- The key is to find a right-angled triangle and then solve it as described above.

> **REMEMBER**
> Always redraw and label the triangle. This avoids confusion with any other numbers that might be on the diagram provided.

For health and safety reasons the top of a ladder should be placed at 15° to the vertical. A ladder is 4 m long. How far from a wall should the ladder be placed?

Step 1: Draw the triangle and identify the relevant information. x is opposite to the angle and 4 m is the hypotenuse.

Step 2: Decide which ratio to use. Only one ratio uses opposite and hypotenuse: **sine.**

Step 3: Set up the calculation. Remember opp = hyp $\times \sin \theta$, so the calculation is $x = 4 \times \sin 15°$.

Step 4: Work it out. $x = 1.04$ metres.

Questions

Grade B

1 Find the angle or length marked x in each of the triangles below.

Grade B

2 The angle of elevation to a church spire is measured as 18° from a point 80 m from the base. Calculate the height of the spire.

More advanced trigonmetry

2-D trigonometric problems

- Trigonometry can be used to solve more complicated problems.

> **REMEMBER**
> Do not round off answers too early as the final answer may be outside the acceptable range. Always use the calculator display or at least 4 sf.

Work out the area of this triangle.

Draw the perpendicular from A to BC.

Label the point where it meets CB, X.

Using triangle CXA, $AX = 8 \times \sin 50° = 6.13$ cm

Area $= \frac{1}{2}bh = \frac{1}{2} \times CB \times AX = \frac{1}{2} \times 12 \times 6.13 = 36.8$ cm²

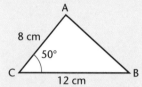 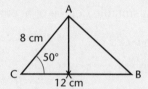

A building has a ledge three-quarters of the way up, along the front, as shown in the diagram (seen from the side).

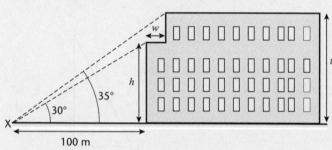

The angle of elevation from a point X, 100 metres away from the building, to the ledge is 30°.

The angle of elevation of the top of the building from the same point is 35°.

Work out: **a** the height of the ledge, h

b the height of the building, t

c the width of the ledge, w

a Using the triangle opposite:
$h = 100 \times \tan 30° = 57.7$ m

b h is three-quarters of t, so $t = 77.0$ m

c Using the triangle opposite:
$x = 77 \div \tan 35° = 110.0$ m

so the width of the ledge is 10 m.

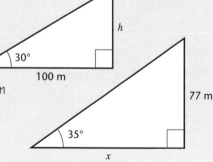

Questions

1 Refer to the second triangle in the first example box above. Work out:

   a CX   **b** XB   **c** angle ABX

PS 2 Refer to the diagram of the building above. The ledge is half the width of the building. An aerial that is 8 metres high is erected at the rear of the building. What is the angle of elevation of the top of the aerial from the point X?

Geometry

3-D trigonometry

3-D trigonometric problems

- On page 46, you revised solving 3-D problems using Pythagoras' theorem.

- You need to follow similar steps when solving 3-D problems using trigonometry.

  - Step 1: Identify the right-angled triangle that includes the required information

  - Step 2: Redraw this triangle as a separate right-angled triangle and label with the given information and the information to be found

  - Step 3: Work out the required value by trigonometry or Pythagoras, or identify another right-angled triangle if necessary

  - Step 4: Solve the problem and round to a suitable degree of accuracy

This cuboid has dimensions 10 cm by 15 cm by 25 cm.
M is the midpoint of FG. Calculate angle AMB.

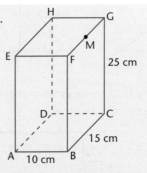

First, identify a right-angled triangle that contains angle AMB and draw it.
This gives △AMB which contains AB = 10 cm
and two lengths that you do not know, MB and AM.

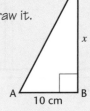

You need to find either MB or AM.

MB will provide an easier calculation, so let MB = x.

Next identify a right-angled triangle that contains MB and draw it.
This gives isosceles triangle BMC.

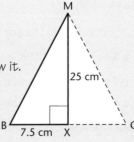

Using Pythagoras' theorem:
$$MB = \sqrt{25^2 + 7.5^2} = \sqrt{681.25} = 26.1 \text{ cm}$$

We can now go back to △AMB and use tangent to work out the angle.
$$AMB = \tan^{-1}(10 \div 26.1) = \tan^{-1} 0.383 = 21°$$

Questions

1 Refer to the cuboid above.

  a Calculate angle AGB.

  b Calculate angle GEC.

PS 2 A triangular prism has a cross-section that is an equilateral triangle with a side of 5 cm and a length of 10 cm. M is the mid-point of the side AB.

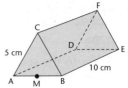

  a Calculate the angle AFB.

  b Calculate the length MF.

Trigonometric ratios of angles from 0 to 360°

AQA 3   EDEXCEL 3   OCR 1

Trigonometric ratios of angles from 0° to 360°

- All the angles that you have revised with sine and cosine up until now have been less than 90°.

- Many triangles have obtuse angles so you will need to be able to find the sine and cosine values of angles greater than 90°.

- All angles, no matter what size, have sine, cosine and tangent values.

- This section revises sine and cosine values of angles from 0° to 360°.

Sine and cosine values of angles from 0° to 360°

- The graphs below shows sin *x* from 0° to 360° and cos *x* from 0° to 360°.

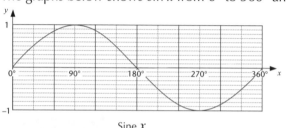

Sine *x*

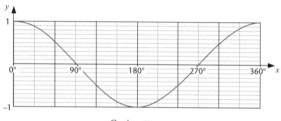

Cosine *x*

- Note the following facts.

  - Each and every value of sine and cosine between –1 and 1 gives two possible angles between 0° and 360°

  - When the value of sine is positive, both angles are between 0° and 180°

  - When the value of sine is negative, both angles are between 180° and 360°

  - When the value of cosine is positive, one angle is between 0° and 90° and the other is between 270° and 360°

  - When the value of cosine is negative, both angles are between 90° and 270°

Find two angles with a sine of –0.2.
You know that both angles are between 180° and 360°.
Use your calculator to find $\sin^{-1} 0.2 = -11.5°$
So the angles are 180° + 11.5° and 360° – 11.5° which gives 191.5° and 348.5°.

> **REMEMBER**
> If you use your calculator to find $\sin^{-1}$ –0.2, you will get a negative answer. Check sin 191.5 and sin 348.5 you get –0.2.

Find the angles with a cosine of –0.8.
You know that both angles are between 90° and 270°.
Use your calculator to find $\cos^{-1} 0.8 = 36.9°$
From the graph, and using symmetry, the angles are:
180° – 36.9° = 143.1°, 180° + 36.9° = 216.9°
So the two angles are 143.1° and 216.9°.

> **REMEMBER**
> Use your calculator to check cos 143.1 and cos 216.9 = –0.8.

Questions

Grade A*

1 State the two angles between 0° and 360° with each of these values.

  a a sine of 0.4   **b** a cosine of 0.1   **c** a sine of –0.7   **d** a cosine of –0.4

Solving any triangle

- Up until now you have only revised right-angled triangles.

- Any triangle has six elements: three sides and three angles.

- To solve a triangle, that is to find any unknown angle or side, you need to know at least three of the elements. Any combination, except for three angles, is enough to work out the rest.

- You can use either the sine rule or the cosine rule to solve a triangle that does not contain a right angle.

The sine rule

- This is the proof of the sine rule.

Take a triangle, ABC, and draw the perpendicular from A to the opposite side BC. Label the point of intersection D and the line h.

From right-angled triangle ADB, $h = c \sin B$

From right-angled triangle ADC, $h = b \sin C$

Therefore, $c \sin B = b \sin C$ which can be rearranged to give $\frac{c}{\sin C} = \frac{b}{\sin b}$

By algebraic symmetry, we see that $\frac{a}{\sin A} = \frac{c}{\sin C}$ and $\frac{a}{\sin A} = \frac{b}{\sin b}$

This can be combined and inverted to give $\frac{\sin A}{a} = \frac{\sin B}{b} = \frac{\sin C}{c}$

Note that the convention is to label angles with capital letters and sides with lower case letters, with the corresponding letters opposite each other.

> **REMEMBER**
> Only the version of the sine rule with the *sides on top* is given on the GCSE formula sheet. If you need to find an angle, remember to invert it.

In triangle ABC, find the value of x.

Use the sine rule with sides on top, which give $\frac{x}{\sin 48°} = \frac{12}{\sin 63°}$

$\Rightarrow x = \frac{12 \sin 48°}{\sin 63°} = 10.0$ cm (3 significant figures)

The ambiguous case

- When you find an angle, you end up with a value of sine, such as $\sin x = 0.9$.

- As you know from page 51, this gives two angles between 0° and 180° giving two possible triangles.

> **REMEMBER**
> Examiners will not try to catch you out with this. They will indicate clearly, either in writing or with the aid of a diagram, what is required.

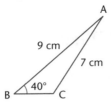

or

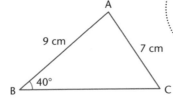

Questions

Grade A

1 Refer to the triangle in the example box above.
  a Find the value of angle B.
  b Use the sine rule to find the length of side AC.

Grade A

2 Refer to the triangle with sides of 9 cm and 7 cm and an angle of 40° above. Use the sine rule to find the two possible values of angle C.

Cosine rule

The cosine rule

- This is the proof of the cosine rule.

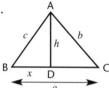

Take a triangle ABC and draw the perpendicular from A to the opposite side BC. Label the point of intersection D and the line AD, h.

Using Pythagoras on triangle BDA, $h^2 = c^2 - x^2$

Using Pythagoras on triangle ADC, $h^2 = b^2 - (a - x)^2$

Therefore, $c^2 - x^2 = b^2 - (a - x)^2$

$$c^2 - x^2 = b^2 - a^2 + 2ax - x^2$$

$$c^2 = b^2 - a^2 + 2ax$$

From triangle BDA, $x = c \cos B$

Hence, $c^2 = b^2 - a^2 + 2ac \cos B$

Rearranging gives $b^2 = a^2 + c^2 - 2ac \cos B$

By algebraic symmetry $a^2 = b^2 + c^2 - 2bc \cos A$ and $c^2 = a^2 + b^2 - 2ab \cos C$

$a^2 = b^2 + c^2 - 2bc \cos A$ can be rearranged to give

$$\cos A = \frac{b^2 + c^2 - a^2}{2bc}$$

> **REMEMBER**
>
> The cosine rule
> $a^2 = b^2 + c^2 - 2bc \cos A$ is given on the formula sheet that is included with the examination but the rearranged formula for the angles is not. You should try to learn this as mistakes are frequently made when trying to rearrange it.

Find x in this triangle.

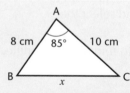

By the cosine rule, $x^2 = 8^2 + 10^2 - 2 \times 8 \times 10 \times \cos 85°$

$$x^2 = 150.1$$

$$x = 12.2 \text{ cm}$$

> **REMEMBER**
>
> Be careful when calculating the cosine rule. A common error is to work out $(8^2 + 10^2 - 2 \times 8 \times 10) \times \cos 85$. Make sure you can use your calculator properly.

Find x in this triangle.

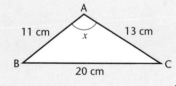

By the cosine rule, $\cos x = \dfrac{11^2 + 13^2 - 20^2}{2 \times 11 \times 13}$

$$\cos x = -0.3846...$$

$$x = 112.6°$$

> **REMEMBER**
>
> If the answer for cosine is negative, the angle will be obtuse.

Questions

Grade A

1 Find the length x in this triangle. Give your answer to 1 decimal place.

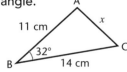

Grade A

2 Find angle x in this triangle.

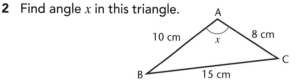

Geometry

Solving triangles

Which rule to use

- When solving triangle problems, there are only four possible situations that can occur, each of which can be solved completely in three stages.

  – **Two sides and the included angle**
    1. Use cosine rule to find the third side
    2. Use the sine rule to find either of the other two angles
    3. Use the sum of the angles in a triangle to find the third angle

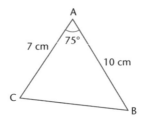

  – **Two angles and a side**
    1. Use the angle sum of the angles in a triangle to find the third angle
    2. Use the sine rule to find either of the other two sides
    3. Use the sine rule to find the third side

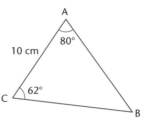

  – **Three sides**
    1. Use the cosine rule to find any one of the angles
    2. Use the sine rule to find either of the other two angles
    3. Use the sum of the angles in a triangle to find the third angle

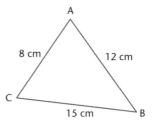

  – **Two sides and a non-included angle**
    1. Use the sine rule to find the two possible values of the appropriate angle. Use information in the question to decide if the angle is acute or obtuse and select the appropriate value.
    2. Use the sum of the angles in a triangle to find the third angle.
    3. Use the sine rule to find the third side.

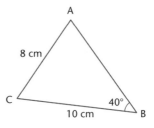

Using sine to find the area of a triangle

- If you know two sides and the included angle of a triangle, you can use the formula:

  area $= \frac{1}{2} \times a \times b \times \sin c$

  to calculate its area.

Find the area of triangle ABC.

Area $= \frac{1}{2} ab \sin C$

Area $= \frac{1}{2} \times 8 \times 10 \times \sin 42 = 26.8$ cm^2

(3 significant figures)

Questions

Grade A

1. a Find the side BC in the first triangle above.
   b Find the side BC in the second triangle above.

Grade A

2. a Find angle A in the third triangle above.
   b Find angle A in the fourth triangle above. It is an obtuse angle.

Grade A

3. Find the area of the first triangle above.

Grade A

AU **4** a Describe the steps required, as in the 'Which rule to use' section above, to calculate the area of this triangle.

   b Use your method to work out the area of the triangle.

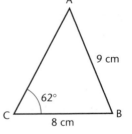

Polygons

Special quadrilaterals

- There are many different quadrilaterals. You will already know the **square** and the **rectangle**.

- The **square** has the following properties:
  - all sides are equal
  - all angles are equal
  - diagonals bisect each other and cross at right angles

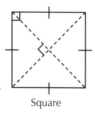
Square

- The **rectangle** has the following properties:
  - opposite sides are equal
  - all angles are equal
  - diagonals bisect each other

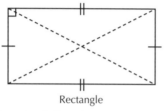
Rectangle

- The **parallelogram** has the following properties:
  - opposite sides are equal and parallel
  - opposite angles are equal
  - diagonals bisect each other

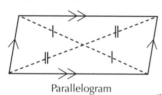
Parallelogram

- The **rhombus** has the following properties:
  - all sides are equal
  - opposite sides are parallel
  - opposite angles equal
  - diagonals bisect each other and cross at right angles

Rhombus

- The **kite** has the following properties:
  - two pairs of sides are equal
  - one pair of opposite angles are equal
  - diagonals cross at right angles

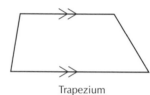
Kite

- The **trapezium** has the following properties:
  - two unequal sides are parallel
  - interior angles add to 180°
  - an **isosceles trapezium** has a **line of symmetry**

Trapezium

REMEMBER

If you are asked about a quadrilateral, draw it and draw in the diagonals. Then mark on all the equal angles and sides.

Regular polygons

- A **regular polygon** is a polygon with all its sides the same length.
  Here are three regular polygons.

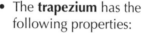

- In any regular polygon, all of the **interior angles**, i, are equal, and all of the **exterior angles**, e, are equal.

- To find the interior angle of any regular polygon, **divide the angle sum** (S) (given by $S = 180(n - 2)°$) by the number of **sides** (n).

- To find the exterior angle of any regular polygon, **divide 360°** by the number of **sides**.

Questions

Grade D–C

AU 1 **a** Marcie says, 'All squares are rectangles.' Is she correct?

  b Milly says, 'All rhombuses are parallelograms.' Is she correct?

  c Molly says, 'All kites are rhombuses.' Is she correct?

  Give a reason for each answer.

Grade C

2 What is **i** the interior angle and **ii** the exterior angle of:

  a a regular octagon

  b a regular nonagon

Grade C

AU 3 What is the connection between the interior and exterior angles of any regular polygon?

Geometry

Circle theorems

Circle theorems

- Here are four circle theorems that you will need to know for your GCSE examination.

  - Angles at the circumference in the same segment of a circle are equal

    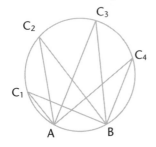

  - Every angle at the circumference of a semicircle that is subtended by the diameter of the semicircle is a right angle

    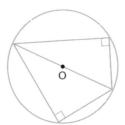

  - The angle at the centre of a circle is twice the angle at the circumference subtended by the same arc

    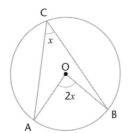

  - The sum of the opposite angles of a cyclic quadrilateral is 180°. (A cyclic quadrilateral is one in which all four verticies lie of the circumference of a circle.)

    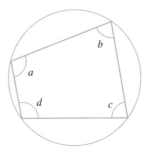

- These theorems can be used to find angles in circle problems.

O is the centre of each circle. Find the angles marked with letters in each of these diagrams. Give a reason for each answer.

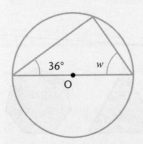 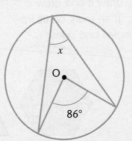 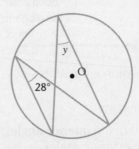 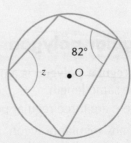

- w is 54° (angles in a semicircle, sum of angles in a triangle)

- x is 43° (angle at the centre = twice the angle at circumference)

- y is 28° (angles in the same segment are equal)

- z is 98° (opposite angles in a cyclic quadrilateral)

Questions

1 O is the centre of the circle.

  a Find the angle marked a.
Give a reason for your answer.

  b Find the angle marked b.
Give a reason for your answer.

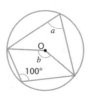

2 O is the centre of the circle.

  a Find the angle marked p.
Give a reason for your answer.

  b Find the angle marked q.
Give a reason for your answer.

Geometry

Tangents and chords

- A **tangent** is a straight line that touches a circle at only one point.

- A **chord** is a line that joins two points on the circumference of a circle.

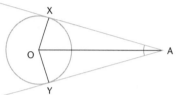

- There are some facts associated with tangents and chords that are useful in solving circle problems.

  – A tangent to a circle is perpendicular to the **radius** drawn to the point of contact

  – Tangents to a circle from an external point to the points of contact are equal in length

  – The line joining an external point to the centre of a circle bisects the angle between the tangents

  – A radius perpendicular to a chord bisects the chord at 90° and when each end of the chord is joined to the centre of the circle, an isosceles triangle is formed

- These facts can be used to find angles in circle problems.

  O is the centre of both circles. Find the angles marked with letters in both of these diagrams. Give reasons for your answers.

  x is 45° (radius is perpendicular to a tangent so there are two 90° angles in the quadrilateral hence $90° + 90° + 3x + x = 360°$)

  y is 124° (the third angle in an isosceles triangle in which the other two angles are 28°)

  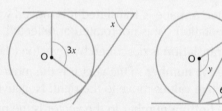

Alternate segment theorem

- The alternate segment theorem states that the angle between a tangent and a chord through the point of contact is equal to the angle in the alternate segment.

  Find the angle marked x in the diagram.

  The angle in the alternate segment is 70°.

  Therefore $x = 180 - 75 - 70 = 35°$.

  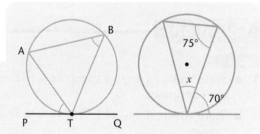

Questions

1 What is the value of x. Give a reason for your answer.

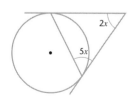

2 What is the value of p. Give a reason for your answer.

Geometry

Transformation geometry

Transformations

- When a shape is **moved**, **rotated**, **reflected** or **enlarged**, it is called a **transformation**.
- There are four transfomations used in GCSE examinations: **translation**, **reflection**, **rotation** and **enlargement**.
- Shapes that are translated, reflected or rotated are **congruent** to the original shape.
- Shapes that are enlarged are **similar** to each other.
- The original shape is called the **object**, and the transformed shape is called the **image**.
- Sometimes the transformation is said to **map** the object onto the image.

Congruent triangles

- Triangles that are exactly the same shape and size are called **congruent** triangles.
- A triangle must meet one or more of the **four conditions** below to be congruent.
  - **SSS** (side, side, side) means all three sides in one triangle are equal to the corresponding sides of the other triangle.
  - **SAS** (side, angle, side) means that two sides and the angle between them in one triangle are equal to the the two corresponding sides and the angle between them in the other triangle.
  - **ASA** (angle, side, angle) means that two angles and a side in one triangle are equal to the corresponding angles and side in the other triangle.
  - **RHS** (right angle, hypotenuse, side) means that both triangles have a right angle, their hypotenuse and another equal side are equal.

REMEMBER

To check or work out a translation, use tracing paper. Trace the shape and then count squares as you move it horizontally and vertically.

Translations

- When a shape is **translated** it is moved without altering its orientation – it is not rotated or reflected.
- A **translation** is described by a vector. $\binom{-4}{5}$ is a vector.
- The **top number** in the vector is the movement in the x-direction. Positive values move to the right. Negative values move to the left.
- The **bottom number** in the vector is the movement in the y-direction. Positive values move upwards. Negative values move downwards.

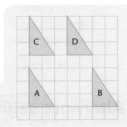

Triangle C is translated from triangle A by the vector $\binom{0}{4}$.
Triangle B is translated to triangle D by the vector $\binom{-2}{4}$.

Questions

Grade B

1 a State which, if any, of triangles A to D below are congruent.

  b State which, if any, of triangles A to D are similar.

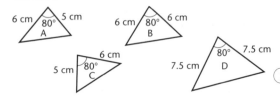

Grade C

2 Refer to the diagram above. What vector translates:

  a triangle A to triangle B

  b triangle A to triangle D

  c triangle C to triangle B

  d triangle C to triangle A

Grade C

AU 3 Are all congruent triangles also similar triangles? Explain your answer.

Reflections

- When a shape is **reflected** it becomes a **mirror** image of itself.

- A **reflection** is described in terms of a mirror line.

- **Equivalent points** on either side are the **same distance** from the mirror line and the **line joining them** crosses the **mirror line** at **right angles**.

REMEMBER
Use tracing paper to check or find the image. The tracing paper can be folded along the mirror line.

Triangle P is a reflection of the shaded triangle in the x-axis.

REMEMBER
The mirror lines in GCSE questions will always be of the form $y = a$, $x = b$, $y = x$, $y = -x$.

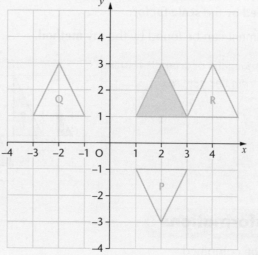

Rotations

REMEMBER
To check or work out a rotation, use tracing paper. Use a pencil point on the centre and rotate the tracing paper in the appropriate direction, through the given angle.

- When a shape is **rotated** it is turned about a centre, called the **centre of rotation**.

- The rotation will either be in a **clockwise** or **anticlockwise** direction.

- The rotation can be described by an **angle**, such as 90°, or a **fraction of a turn**, such as 'a half-turn'.

Triangle A is a rotation of 90° of the shaded triangle in a clockwise direction about the centre (1, 0).

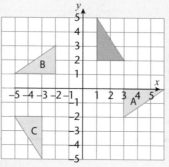

- A rotation of 90° clockwise is the same as a rotation of 270° anticlockwise. Only a half-turn does not need to have a direction specified.

Questions

Grade D

1 Refer to the diagram in the first example box. What is the mirror line for the reflection:

   a that takes the shaded triangle to triangle Q?

   b that takes the shaded triangle to triangle R?

Grade D

2 Refer to the diagram in the second example box. What is the rotation that takes:

   a the shaded triangle to triangle B?

   b the shaded triangle to triangle C?

Enlargements

C

- When a shape is **enlarged**, it changes its size to become a shape that is **similar** to the first shape.

- An **enlargement** is described by a **centre of enlargement** and a **scale factor**.

- The lengths of the sides of the image will be the lengths of the sides of the object **multiplied** by the **scale factor**.

- The centre of enlargement can be found by the **ray method**.

> **REMEMBER**
> Scale factors can also be fractions. In this case the image is smaller than the original object.

Triangle Q is an enlargement of triangle P with a scale factor of 3 from the centre O.

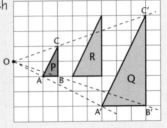

Combined transformations

B

- Transformations can be combined.

Triangle A is a rotation of the shaded triangle 90° clockwise about (0, 1).

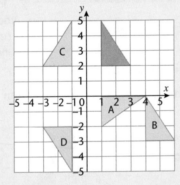

Triangle B is a rotation of triangle A 90° anticlockwise about (4, 0).

To get back to the shaded triangle, triangle B is translated by the vector $\binom{-3}{5}$.

Questions

Grade C

1 Refer to the diagram in the first example box.

  a i Write down the scale factor of the enlargement that takes triangle P to triangle R.

    ii Mark the centre of the enlargement that takes triangle P to triangle R.

  b i Write down the scale factor of the enlargement that takes triangle Q to triangle R.

    ii Mark the centre of the enlargement that takes triangle Q to triangle R.

Grade B

2 Refer to the diagram in the second example box.

  a Describe fully the transformation that takes the shaded triangle to triangle C.

  b Describe fully the transformation that takes triangle C to triangle D.

  c What **single** transformation takes triangle D to the shaded triangle?

Constructions

D

Constructing triangles

- When you are asked to **construct** a triangle, you are expected to use a ruler with a pair of **compasses** to measure lengths and a **protractor** to measure angles.

- There are **three ways** of constructing triangles.

- **All three sides given**

  - Use a ruler to draw one side. (Sometimes this side is already drawn.)

  - Use the ruler to set the compasses to the length of each of the other two sides, in turn, and draw arcs from the ends of the side you have drawn.

  - Join up the ends of the line to the point where the arcs intersect.

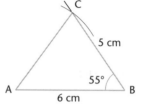

- **Two sides and the included angle given**

  - Use a ruler to draw one side. (Sometimes this side is already drawn.)

  - Use a protractor to measure and draw the angle at one end of the line you have drawn.

  - Set the compasses to the length of the other given side and draw an arc, cutting the line you have just drawn for the angle.

  - Join up the points where the arc cuts the line for the angle to the other end of the base line.

> **REMEMBER**
> These questions are often not answered well in examinations because candidates forget the techniques. The night before the exam, look up 'Euclidean constructions' on the internet to find websites that will give you a dynamic demonstration of all the constructions you need for GCSE.

- **Two angles and the side between them given**

  - Use a ruler to draw the side. (Sometimes this side is already drawn.)

  - Use a protractor to measure and draw the angles at each end of the line.

  - Extend the lines to form the triangle.

Constructing an angle of 60°

- You can use **compasses** and a **ruler** to draw an angle of 60° accurately.

  - Draw a line and mark a point where the angle will be drawn.

  - With the compasses centred on this point, draw an arc that cuts the line.

  - With the compasses set to the same radius, and centred on the point where the first arc cuts the line, draw another arc to cut the first arc.

  - Join the original point to the point where the arcs cross.

> **REMEMBER**
> Always show your construction lines and arcs clearly. You will not get any marks if they can't be seen.

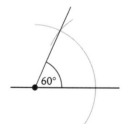

C

Questions

Use a ruler, compasses and protractor for these questions.

Grade D

1 a Draw accurately the triangle with sides of 4 cm, 5 cm and 6 cm, shown above.

  b Draw accurately the triangle with sides of 6 cm and 5 cm and an included angle of 55°, shown above.

c Draw accurately the triangle with a side of 7 cm, and angles of 40° and 65° shown above.

Grade C

2 Follow the steps shown above and construct an angle of 60°.

Use a protractor to check the accuracy of your drawing.

AQA 3   EDEXCEL 3   OCR 1

The perpendicular bisector

- To **bisect** means to **divide in half**.

- **Perpendicular** means **at right angles**.

- A **perpendicular bisector** divides a line in two, and is at right angles to it.

    - Start off with a line or two points. (Usually, these are given.)

    - Open the compasses to a radius about three-quarters of the length of the line, or of the distance between the points.

    - With the compasses centred on each end of the line (or each point) in turn, draw arcs on both sides of the line.

    - Join up the points where the arcs cross.

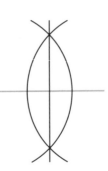

The angle bisector

- An **angle bisector** divides an angle into two smaller, equal angles.

    - Start off with an angle. (Usually, this is given.)

    - Open the compasses to any radius shorter than the arms of the angle. With the compasses centred on the vertex of the angle (the point where the arms meet) draw an arc on both arms of the angle.

    - Now, with the compasses still set to the same radius and centred on the points where these arcs cross the arms, draw intersecting arcs.

    - Join up the vertex of the angle and the point where the arcs cross.

The perpendicular at a point on a line

- The perpendicular at a point on a line is a line at right angles to the line, passing through the point.

    - Start off with a point on a line. (Usually, this is given.)

    - Open the compasses to a radius of about 3 cm and centred on the point, draw arcs on either side.

    - Now increase the radius of the compasses to about 5 cm and draw arcs centred on the two points on either side, so that they intersect, cutting the line.

    - Join the original point on the line and the point where the arcs cross.

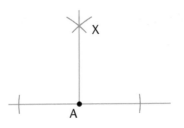

Questions

Grade C

1 Draw a line 6 cm long.

Following the steps above, draw the perpendicular bisector of the line.

Check that each side is 3 cm long and that the angle is 90°.

Grade C

2 Draw an angle of 70°.

Following the steps above, draw the angle bisector.

Check that each half-angle is 35°.

Grade C

3 Draw a line and mark a point on it.

Following the steps above, draw the perpendicular at the point to the line. Check that the angle is 90°.

Geometry

Constructions and loci

AQA 3   EDEXCEL 3   OCR 1

The perpendicular from a point to a line

- The **perpendicular** is a line that is at right angles to the original line and passes through the point.

  - Start with a line and a point not on the line. (Usually, these are given.)

  - Open the compasses to about 3 cm more than the distance from the point to the line. With the compasses centred on the point, draw arcs on the line on either side of the point.

  - Centring the compasses on the points where these arcs cut the line, draw arcs on the other side of the line so they intersect.

  - Join the original point and the point where the arcs cross.

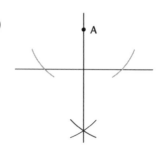

Loci

- A **locus** (plural **loci**) is the path followed by a point according to a rule.

| A point that moves so that it is always 4 cm from a fixed point is a circle of radius 4 cm. | A point that moves so that it is always the same distance from two fixed points is the perpendicular bisector of the two points. |

Practical problems

- **Loci** can be used to solve real-life problems.

A horse is tethered to a rope 10 m long in a large, flat field. What is the area the horse can graze?

The horse will be able to graze anywhere within a circle of radius 10 m.

- In reality the horse may not be able to graze an exact circle but the situation is modelled by the mathematics.

Questions

Grade C

1 Following the steps above, draw the perpendicular from a point to a line. Check that the angle is 90°.

Grade C

2 A radar station in Edinburgh has a range of 200 miles.
A radar station in London has a range of 250 miles.
London and Edinburgh are 400 miles apart.
Sketch the area that the radar stations can cover.
Use a scale of 1 cm = 200 miles.

Similarity

Similar triangles

- Two triangles are **similar** if their corresponding angles are equal.
- The corresponding sides of similar triangles are in the **same ratio**.
- The **ratio** between the lengths of corresponding sides is called the **scale factor**.

> **REMEMBER**
> Exam questions often try to catch you out by asking about the size of angles. Angles are the same in both triangles.

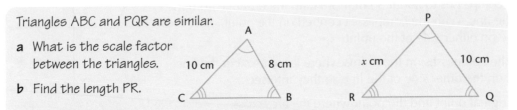

Triangles ABC and PQR are similar.

a What is the scale factor between the triangles.

b Find the length PR.

a Comparing sides AB and PQ we can see that the scale factor is $10 \div 8 = 1.25$

b PR is, therefore, $10 \times 1.25 = 12.5$ cm

Special cases of similar triangles

- Some special cases of similar triangles frequently appear in GCSE examinations.
- The first is when two triangles share a common angle and parts of common sides.
- The key to solving this type of problem is to redraw the two triangles separately.

> **REMEMBER**
> Put the unknown on the top when you set up the equations, then cross-multiply as in the example below.

Find the sides marked x and y in these triangles.

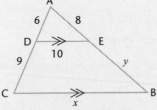
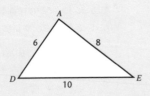
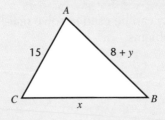

Redraw the triangles.

Triangles AED and ABC are similar. So using the corresponding sides CB, DE with AC, AD $\dfrac{x}{10} = \dfrac{15}{6} \Rightarrow x = 10 \times \dfrac{15}{6} = 25$

Using the corresponding sides AB, AE with AC, AD gives

$\dfrac{8 + y}{8} = \dfrac{15}{6} \Rightarrow 8 + y = 8 \times \dfrac{15}{6} = 20 \Rightarrow y = 12$

Questions

Grade B

1 To work out the height of a church, Zoe stands a 2 metre stick on the ground then lines up a point on the ground with the top of the stick and the top of the spire. She then paces out the distances.

Use similar triangles to work out the height of the spire.

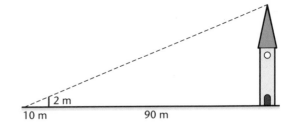

Geometry

Similar shapes

- When two shapes are similar, there are relationships between their lengths, areas and volumes.

- The corresponding **lengths** of similar shapes are in the same ratio, the **length ratio**, which is equal to the **linear scale factor**.

- The **areas** of similar shapes are in the same ratio, called the **area ratio** or **area scale factor**. This is equal to the **square** of the length ratio.

- The **volumes** of similar 3-D shapes are also in the same ratio, called the **volume ratio** or **volume scale factor**. This is equal to the **cube** of the length ratio.

- If the linear scale factor is n, the area scale factor is n^2 and the volume scale factor is n^3.

> **REMEMBER**
> The key word in identifying these problems is **similar**, followed by a reference to either **area** or **volume**.

Every length in cuboid B is twice the corresponding length in cuboid A.

The length of A is 5 cm and the length of B is 10 cm, so the lengths are in the ratio $5 : 10 = 1 : 2$.

The area of the front face of A is 15 cm² and the area of the front face of B is 60 cm², so the ratio of the areas is $15 : 60 = 1 : 4 = 1 : 2^2$.

The volume of cuboid A is 30 cm³ and the volume of cuboid B is 240 cm³, so the ratio of the volumes is $30 : 240 = 1 : 8 = 1 : 2^3$.

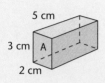
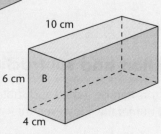

Solving problems with area and volume ratios

- It is essential to follow three steps when solving problems about similar shapes

    – Step 1: Establish the given scale factor

    – Step 2: Establish the scale factor(s) required to solve the problem

    – Step 3: Multiply or divide the given dimension by the appropriate scale factor to find the required dimension

> **REMEMBER**
> If you are given the volume or area scale factor and you need to find the other of these. Find the linear scale factor first as this makes working easier.

Soup is sold in similar small and large cans. The volume of the large can is 500 ml and the volume of the small can 300 ml.

The labels on the cans are also similar in shape. The label on the large can has an area of 40 cm². What is the area of the label on the small can?

Step 1: The volume scale factor is $500 \div 300 = 1.\dot{6}$

Step 2: The linear scale factor $= \sqrt[3]{1.\dot{6}}$ and the area scale factor $= (\sqrt[3]{1.\dot{6}})^2$

Step 3: Let the area of the label of the small can be A, then $A = 40 \div (\sqrt[3]{1.\dot{6}})^2 = 28.46$ cm².

Questions

1 A cuboid is 3 cm by 8 cm by 12 cm. A similar cuboid is 9 cm by 24 cm by 36 cm.

 a What is the linear scale factor?

 b What is the area scale factor?

 c What is the volume scale factor?

PS 2 Two similar statuettes are made from the same material. They are 15 cm and 25 cm tall respectively. The smaller statuette has a mass of 4 kg. If mass is proportional to volume, calculate the mass of the larger statuette.

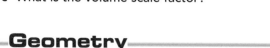

Vectors

A

Vectors

- A **vector** is a quantity that has both **magnitude** and **direction**.

- A vector can be represented by a straight line with an arrowhead pointing in the direction of the vector and the length representing the magnitude of the vector.

- Vectors are written as single letters in bold type, for example **a**, or as a pair of letters with an arrow above, for example $\vec{PQ}$.

The vector **p** shows a translation from A to B in the direction from A to B.

It can also be written as $\vec{AB}$.

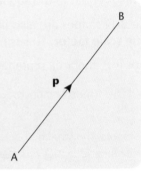

Addition and subtraction of vectors

A

- Vectors can be added and subtracted, using the normal rules of algebra and numeracy. However, they also represent movements in space.

a + **b** and **a** − **b** are shown in the diagrams opposite.

REMEMBER

If $\vec{OA}$ = **a** and $\vec{OB}$ = **b**, then $\vec{AB}$ = **b** − **a**.

- When vectors relate to an origin, they are called **position vectors**.

In the grid below $\vec{OA}$ = **a**,

$\vec{OB}$ = **b**,

$\vec{OP}$ = 3**a** + 2**b**,

$\vec{OM}$ = **a** + 3**b**

$\vec{HK}$ = 2**a** − **b**,

$\vec{RA}$ = −3**a** − 2**b**,

and $\vec{IH}$ = **a** − **b**.

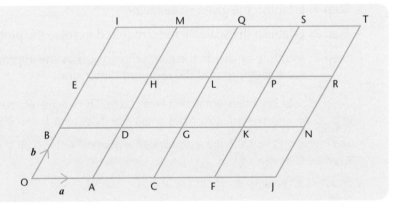

Questions

1 Refer to the grid above.

   a Write down the following vectors in terms of **a** and **b**.

   i $\vec{EM}$ **ii** $\vec{DN}$ **iii** $\vec{DS}$

   iv $\vec{KH}$ **v** $\vec{RF}$

   b What is the relationship between the vectors $\vec{BH}$ and $\vec{AT}$?

   c What is the relationship between the vectors $\vec{BQ}$ and $\vec{RC}$?

   AU d What does the fact that $\vec{OD}$ = **a** + **b** and $\vec{DS}$ = 2**a** + 2**b** tell you about the points O, D and S?

Geometry

Vector geometry

- **Vectors** can be used to prove results in geometry.

- In vector geometry problems, the key is to find a 'route' following vectors that you know or can find.

- When they form a 'route', vectors link together so that the last letter of one vector will be the first letter of the second vector.

  e.g. $\vec{AB} = \vec{AX} + \vec{XY} + \vec{YZ} + \vec{ZB}$

- GCSE questions usually ask for a conclusion to be made. This consists of three possible answers.

  - The vectors are parallel. In this case the vectors are multiples of each other.

  - The vectors are on a straight line. In this case the vectors are multiples of each other and have a common point.

  - The vectors are opposite to each other. In this case the vectors will be negative multiples of each other.

- The opposite (negative) of any vector can be formed by reversing the letters.
  $\vec{AB} = -\vec{BA}$

> **REMEMBER**
> GCSE questions always start with finding simple vectors which are needed later for more complex vectors. You should always be aware of using earlier answers.

A*

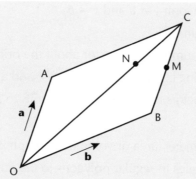

OACB is a parallelogram.
$\vec{OA} = \mathbf{a}$, $\vec{OB} = \mathbf{b}$, M is the midpoint of BC and N is the point such that
ON:NC = 2:1.

Show that ANM is a straight line.
The vector $\vec{ON}$ is $\frac{2}{3}\vec{OC} = \frac{2}{3}(\mathbf{a} + \mathbf{b})$

$\vec{AN} = \vec{AO} + \vec{ON} = -\mathbf{a} + \frac{2}{3}(\mathbf{a} + \mathbf{b}) = -\frac{1}{3}\mathbf{a} + \frac{2}{3}\mathbf{b}$

$\vec{AM} = \vec{AO} + \vec{OB} + \vec{BM} = -\mathbf{a} + \mathbf{b} + \frac{1}{2}\mathbf{a} = -\frac{1}{2}\mathbf{a} + \mathbf{b}$

$\vec{AM} = \frac{3}{2}\vec{AN}$ hence $\vec{AM}$ and $\vec{AN}$ are multiples of each other and share a common point, A, so ANM is a straight line.

Questions

Grade A*

1 In the diagram above, the line OB is extended to a point P such that $\vec{OP} = 2\mathbf{b}$.

   a Write down the following vectors in terms of **a** and/or **b**.

   i $\vec{AB}$ ii $\vec{NC}$ iii $\vec{AM}$

(AU b) Show that AMP is a straight line.

Geometry

Geometry grade booster

I can...

- [ ] find the area of a parallelogram, using the formula $A = bh$
- [ ] find the area of a trapezium, using the formula $\frac{1}{2}(a + b)h$
- [ ] find the area of a compound shape
- [ ] work out the formula for the perimeter, area or volume of simple shapes
- [ ] identify the planes of symmetry for 3-D shapes
- [ ] recognise and find alternate angles in parallel lines and a transversal
- [ ] recognise and find corresponding angles in parallel lines and a transversal
- [ ] recognise and find interior angles in parallel lines and a transversal
- [ ] use and recognise the properties of quadrilaterals
- [ ] find the exterior and interior angles of regular polygons
- [ ] understand the words 'sector' and 'segment' when used with circles
- [ ] calculate the circumference of a circle, giving the answer in terms of π if necessary
- [ ] calculate the area of a circle, giving the answer in terms of π if necessary
- [ ] recognise plan and elevation from isometric and other 3-D drawings
- [ ] translate a 2-D shape
- [ ] reflect a 2-D shape in lines of the form $y = a$ and $x = b$
- [ ] rotate a 2-D shape about the origin
- [ ] enlarge a 2-D shape by a whole number scale factor about the origin
- [ ] construct diagrams accurately using compasses, a protractor and a straight edge
- [ ] use the appropriate conversion factors to change imperial units to metric units and vice versa

You are working at ⬤ Grade D ⬤ level.

- [ ] work out the formula for the perimeter, area or volume of complex shapes
- [ ] relate the exterior and interior angles in regular polygons to the number of sides
- [ ] find the area and perimeter of semicircles
- [ ] translate a 2-D shape, using a vector
- [ ] reflect a 2-D shape in the lines $y = x$ and $y = -x$
- [ ] rotate a 2-D shape about any point
- [ ] enlarge a 2-D shape by a fractional scale factor
- [ ] enlarge a 2-D shape about any centre
- [ ] construct perpendicular and angle bisectors
- [ ] construct an angle of 60°
- [ ] construct the perpendicular to a line from a point on the line and from a point to a line
- [ ] draw simple loci
- [ ] work out the surface area and volume of a prism
- [ ] work out the volume of a cylinder, using the formula $V = \pi r^2 h$
- [ ] find the density of a 3-D shape
- [ ] find the hypotenuse of a right-angled triangle, using Pythagoras' theorem
- [ ] find the short side of a right-angled triangle, using Pythagoras' theorem

☐ use Pythagoras' theorem to solve real-life problems

You are working at **Grade C** level.

☐ use trigonometric ratios to find angles and sides in right-angled triangles

☐ use trigonometry to solve real-life problems involving right-angled triangles

☐ use circle theorems to find angles, in circle problems

☐ use the conditions for congruency to identify congruent triangles

☐ enlarge a shape by a negative scale factor

☐ use similar triangles to find missing lengths

☐ understand simple proofs such as the exterior angle of a triangle is equal to the sum of the opposite interior angles

You are working at **Grade B** level.

☐ calculate the length of an arc and the area of a sector

☐ calculate the surface area of cylinders, cones and spheres

☐ calculate the volume of cones and spheres

☐ solve 3-D problems using Pythagoras' theorem

☐ use the alternate segment theorem to find angles in circle problems

☐ prove that two triangles are congruent

☐ solve more complex loci problems

☐ solve problems using area and volume scale factors

☐ solve real-life problems, using similar triangles

☐ solve problems using addition and subtraction of vectors

☐ use the sine rule to solve non right-angled triangles

☐ use the cosine rule to solve non right-angled triangles

☐ use the rule $A = \frac{1}{2}ab\sin C$ to find the area of non right-angled triangles

You are working at **Grade A** level.

☐ calculate the volume and surface area of compound 3-D shapes

☐ use circle theorems to prove geometrical results

☐ solve more complex problems, using the proportionality of area and volume scale factors

☐ solve problems using vector geometry

☐ use the sine rule to solve more complex problems, involving right-angled and non right-angled triangles

☐ use the cosine rule to solve more complex problems, involving right-angled and non right-angled triangles

☐ solve 3-D problems using trigonometry

☐ find two angles between 0° and 360° for values of sine and cosine

☐ solve simple trigonometric equations where sine or cosine is the subject

☐ prove geometrical results with a logical and rigorous argument

You are working at **Grade A*** level.

Basic algebra

D

Substitution

- Substitution means replacing letters in formulae and expressions with numbers.
- When replacing letters with numbers, use brackets to avoid problems with minus signs.

  Work out the value of $ab + c$ if $a = -3$, $b = 4$ and $c = 5$.
  $ab + c = (-3)(4) + (5) = -12 + 5 = -7$

> **REMEMBER**
> Always use brackets and remember the rules for dealing with negative numbers.

C

Expansion and simplification

- **Expand** in mathematics means **multiply out**.

  Expressions such as $4(z + 3)$ and $5x^2(x - 8)$ can be multiplied out.

- There is an **invisible multiplication sign** between the outside term and the opening bracket.

  $4(2x + 3)$ means $4 \times (2x + 3)$

- When expanding **brackets**, it is important to remember that the term outside the bracket is multiplied by each term inside the brackets.

  $4(2x + 3)$ means $4 \times (2x + 3) = 4 \times 2x + 4 \times 3 = 8x + 12$.
  You would normally just write $5x^2(x - 8) = 5x^3 - 40x^2$.

> **REMEMBER**
> There is no need to show all the steps when expanding brackets as, usually, one mark is given for each correct term.

- When you are asked to **expand and simplify** an expression, it means expand any brackets and then simplify by collecting like terms.

  Expand and simplify $4(3 + m) - 5(2 - 3m)$.
  First, expand both brackets: $12 + 4m - 10 + 15m$
  Second, simplify: $2 + 19m$

> **REMEMBER**
> Do not try to expand and simplify in one go. If you try to do two things at once you will probably do one of them wrong.

C

Factorisation

- **Factorisation** is the opposite of expansion. Factorisation puts an expression back into the form $4(3x - 2)$.

- To factorise expressions, look for the **highest common factor** of both terms.

  $5x + 20$ has a common factor of 5 in each term,
  so $5x + 20 = 5 \times x + 5 \times 4 = 5(x + 4)$
  $4xy - 8x^2$ has a common factor of $4x$, so $4xy - 8x^2 = 4x \times y - 4x \times 2x = 4x(y - 2x)$

- Check your factorisation by multiplying out the final answer to check it goes back to what you started with.

Questions

Grade D

1 If $a = 3$, $b = -4$ and $c = 5$, find the value of each the following.

  a $ab + c$ **b** $a^2 + b^2$ **c** $2(a + 3b - c)$

Grade D

2 Expand the following expressions.

  a $3(x + 5)$ **b** $n(n - 7)$ **c** $3p^2(2p - 3q)$

Grade C

3 Expand and simplify the following expressions.

  a $2(x - 3) + 4(x + 3)$ **b** $8(x + 3y) + 2(x + 7y)$

Grade D

4 Factorise the following expressions.

  a $6x^2 - 9x$ **b** $2a^2b - 8ab + 6ab^2$

Linear equations

AQA 2/3   EDEXCEL 3   OCR 1/2

Solving linear equations

- An **equation** is formed when an expression is set equal to a number or another expression.
- A **linear equation** is one that only involves one **variable**.

  $2x + 3 = 7$ and $5x + 8 = 3x - 2$ are both linear equations.

- **Solving** an equation means finding the value of the variable that makes it true.

  Solve $2x + 3 = 7$. The value of x that makes this true is 2 because $2 \times 2 + 3 = 7$.

- There are four ways to solve equations. There is not much difference between them but **rearrangement** is the most efficient method.

  Solve $6x + 5 = 14$.
  Move the 5 across the equals sign to give: $6x = 14 - 5 = 9$
  Divide both sides by 6 to give: $x = \frac{9}{6} \Rightarrow x = 1.5$

  REMEMBER
  This is called 'change sides, change signs', which means that plus becomes minus (and vice versa) and multiplication becomes division (and vice versa).

Solving equations with brackets

- When an equation contains **brackets** you should multiply out the brackets first and then solve the equation in the normal way.

  Solve $4(x + 3) = 30$.
  Multiply out the brackets to give: $4x + 12 = 30$
  Move the 12 across the equals sign to give: $4x = 30 - 12 = 18$
  Divide both sides by 4 to give: $x = 4.5$

  REMEMBER
  Always check your answer works in the original equation.

Equations with the variable on both sides of the equals sign

- When a letter appears on **both sides** of an equation, use the 'change sides, change signs' rule.
- Use the rule to collect all the terms containing the variable on the left-hand side of the equals signs and all number terms on the right-hand side.

  Rearranging $5x - 3 = 2x + 12$ gives $5x - 2x = 12 + 3$.

  REMEMBER
  Be careful when moving terms across the equals sign and remember to change the signs from plus to minus and vice versa.

- After the equation is rearranged, the terms are collected together and the equation is solved in the usual way.

  $5x - 3 = 2x + 12$ is rearranged to give $5x - 2x = 12 + 3 \Rightarrow 3x = 15 \Rightarrow x = 5$

Questions

Grade C

1 Solve these equations.
  a $3x - 5 = 4$ **b** $2m + 8 = 6$ **c** $\frac{n}{3} = 4$

  d $8x + 5 = 9$ **e** $\frac{x-3}{5} = 2$ **f** $5x - 3 = 7$

  g $\frac{y+2}{5} = 3$ **h** $\frac{x}{7} - 3 = 1$ **i** $8 - 2x = 7$

Grade D

2 Solve these equations.
  a $3(x - 5) = 12$ **b** $2(m - 3) = 6$ **c** $3(x + 5) = 9$
  d $2(x - 3) = 2$ **e** $6(x - 1) = 9$ **f** $4(y + 2) = 2$

Grade D

3 Solve these equations.
  a $3x - 2 = x + 12$ **b** $5y + 5 = 2y + 11$
  c $8x - 1 = 5x + 8$ **d** $7x - 3 = 2x - 8$
  e $6x - 2 = x - 1$ **f** $3y + 7 = y + 2$

Algebra

73

Equations with brackets and the variable on both sides

- When an equation contains brackets and variables on **both sides**, always expand the brackets first.

  Expanding $4(x - 2) = 2(x + 3)$ gives $4x - 8 = 2x + 6$.

- After the brackets are expanded, the equation is rearranged, the terms are collected and the equation is solved in the usual way.

  Expanding $4(x - 2) = 2(x + 3)$ gives $4x - 8 = 2x + 6 \Rightarrow 4x - 2x = 8 + 6 \Rightarrow 2x = 14 \Rightarrow x = 7$.

Setting up equations

- Many **real-life** problems can be solved by using equations to model them.

The angles in a triangle are given by $2x$, $3x$ and $4x$. What is the largest angle in the triangle?

You know the angles in a triangle add up to 180°, so:

$2x + 3x + 4x = 180$

This simplifies to: $9x = 180°$

Dividing both sides by 9 gives: $x = 20$

So the largest angle is $4x = 4 \times 20 = 80°$.

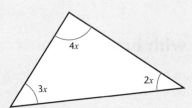

REMEMBER

Use the letter x when setting up an equation unless you are given another letter to use.

Fred is 29 years older than his daughter, Freda. Together their ages add up to 47. Using the letter x to represent Freda's age, write down an expression for Fred's age. Set up an equation in x and solve it to find Freda's age.

Fred is $x + 29$, $x + x + 29 = 47 \Rightarrow 2x + 29 = 47 \Rightarrow 2x = 18 \Rightarrow x = 9$

So Freda is 9 years old.

Questions

Grade C

1 Solve each of the following.

  a $5(x - 3) = 2(x + 6)$ **b** $6(x + 1) = 2(x - 3)$

  c $3(x + 5) = 2(x + 10)$ **d** $7(x - 2) = 3(x + 4)$

Grade D

2 The diagram shows a rectangle.

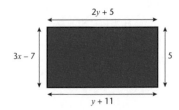

  a What is the value of x?

  b What is the value of y?

Grade C

PS 3 A family has x bottles of milk delivered every day from Monday to Friday and 7 bottles delivered on Saturday. There is no delivery on Sunday.

In total they have 22 bottles delivered each week.

Find the value x.

Grade C

4 Asil thought of a number. He divided it by 2 then added 7. The result was 6 more than the number he first thought of.

  a Use the above information to set up an equation, using x to represent the number Asil thought of.

  b Solve your equation to find the value of x.

Algebra

Trial and improvement

Trial and improvement

- Some equations **cannot be solved** simply by using algebraic methods.

- A **numerical method** for solving these equations is called **trial and improvement**.

- After an initial guess (**the trial**), an answer is calculated and compared to the required answer.

- A better guess (**the improvement**) is then made.

- This process is **repeated** until the answer is within a given accuracy.

- Normally the first thing to do is find **two whole numbers** between which the answer lies.

- Then find **two decimal numbers**, **each with one decimal place**, between which the answer lies.

- Finally, test the **mid-way value** to see which of the two numbers is closer.

> **REMEMBER**
> In GCSE examinations, one or two initial guesses are always given to give you a start.

Use trial and improvement to solve the equation $x^3 + 2x = 52$, giving your answer to 1 decimal place.

| Try $x = 4$ | $4^3 + 2 \times 4 = 72$ | Too high – next trial needs to be smaller |

| Try $x = 3$ | $3^3 + 2 \times 3 = 33$ | Too low |

So we now know that the solution lies between $x = 3$ and $x = 4$.

| Try $x = 3.5$ | $3.5^3 + 2 \times 3.5 = 49.875$ | Close, but too low |

| Try $x = 3.6$ | $3.6^3 + 2 \times 3.6 = 53.856$ | Close, but too high |

So we know the solution lies between $x = 3.5$ and $x = 3.6$.

To see which is closest, try $x = 3.55$ $3.55^3 + 2 \times 3.55 = 51.838\ 875$, which is just too low.

Hence, the answer, correct to one decimal place, is 3.6.

- You could set out the working in a table.

> **REMEMBER**
> You must test the value that is halfway between the one-decimal place values.

Guess	$x^3 + 2x$	Comment
4	72	Too high
3	33	Too low
3.5	49.875	Too low
3.6	53.856	Too high
3.55	51.838 875	Too low

Questions

Grade C

1 Use trial and improvement to find the solution to $x^3 - 2x = 100$.

Give your answer to 1 decimal place.

Grade C

2 Use trial and improvement to find the solution to $x^3 + x = 20$.

Give your answer to 1 decimal place.

Algebra

Simultaneous equations

Simultaneous equations

- **Simultaneous equations** are two or more equations for which we want to find the same solution. These are normally linear equations involving **two variables**.

> $2x + y = 10$ has many solutions: $x = 4, y = 2;$ $x = 3, y = 4; ...$
> and $3x - y = 5$ has many solutions: $x = 4, y = 7;$ $x = 3, y = 4; ...$
> Only one solution, $x = 3$ and $y = 4$, satisfies both equations at the same time.

- Simultaneous equations can be solved by one of two methods:
  - the **elimination** method
  - the **substitution** method

> **REMEMBER**
> Always label the equations so your method can be explained.

- Follow these steps to use the elimination method to solve simultaneous equations.
  - Step 1: **Balance** the **coefficients** of one of the **variables**
  - Step 2: Add or subtract the equations to **eliminate** this variable
  - Step 3: **Solve** the resulting linear equation to find the value of the other variable
  - Step 4: **Substitute** the value back into one of the previous equations
  - Step 5: **Solve** this equation find the value of the second variable
  - Step 6: **Check** your values in the initial equations

> Solve the equations $4x + 3y = 27$ and $5x - 2y = 5$.
> $4x + 3y = 27$ (1)
> $5x - 2y = 5$ (2)
> Balance the coefficent of x by multiplying (1) by 5 and (2) by 4:
> (1) $\times$ 5 $20x + 15y = 135$ (3)
> (2) $\times$ 4 $20x - 8y = 20$ (4)
> Subtract equation (4) from equation (3):
> (3) − (4) $23y = 115 \Rightarrow y = 5$
> Substitute $y = 5$ into (1):   Check: $4 \times 3 + 3 \times 5 = 27$ ✓
> $4x + 15 = 27 \Rightarrow 4x = 12 \Rightarrow x = 3$ $5 \times 3 - 2 \times 5 = 5$ ✓

> **REMEMBER**
> If the balanced terms have the same sign subtract the equations. If they have different signs add the equations.

- The **substitution method** works by substituting one equation into the other.

> Solve the equations $6x + y = 9$ and $y = 3x - 9$.
> $6x + y = 9$ (1)
> $y = 3x - 9$ (2)
> Substitute the right-hand side of equation (2) into equation (1):
> $6x + (3x - 9) = 9$
> Expand and solve the equation:
> $9x - 9 = 9 \Rightarrow 9x = 18 \Rightarrow x = 2$
> Substitute $x = 2$ into equation (1):   Check: $6 \times 2 - 3 = 9$ ✓
> $6 \times 2 + y = 9 \Rightarrow y = -3$ $-3 = 3 \times 2 - 9$ ✓

> **REMEMBER**
> Always test the values in the original equations.

Questions

Grade B

1 Solve these simultaneous equations.

a $3x + y = 7$ **b** $3x + 5y = 15$ **c** $2x + 5y = 15$
   $2x + y = 4$ $x + 3y = 7$ $3x - 2y = 13$

Grade B

2 Solve the equations $4x - y = 17$ and $x = 5 + y$.

Grade B

AU 3 Explain why the simultaneous equations
$3x + 5y = 10$ and $3x + 5y = 15$ have no solution.

Simultaneous equations and formulae

Setting up simultaneous equations

- Many real-life situations can be solved by simultaneous equations.

Four cakes and three cups of tea cost £6.40.

Three cakes and two cups of tea cost £4.50.

How much will it cost to buy three cakes and four cups of tea?

Let x be the cost of a cake.

Let y be the cost of a cup of tea.

$$4x + 3y = 640 \quad (1)$$
$$3x + 2y = 450 \quad (2)$$

Solve these equations using the method already demonstrated:

(1) × 3 $12x + 9y = 1920$ (3)

(2) × 4 $12x + 8y = 1800$ (4)

(3) − (4) $y = 120$

Substitute into (1): $4x + 360 = 640 \Rightarrow 4x = 280 \Rightarrow x = 70$

Check with the original information: $4 \times 70 + 3 \times 120 = 640$, $3 \times 70 + 2 \times 120 = 450$ ✓

The cost of three cakes and four cups of tea will be $3 \times 70 + 4 \times 120 = £6.90$.

Rearranging formulae

- The subject of a formula is the **variable** (letter) in the formula that stands on its own, usually on the left-hand side of the equals sign.

x is the subject of each of these formulae.

$$x = 5t + 4 \qquad x = 4(2y - 7) \qquad x = \frac{1}{t}$$

> **REMEMBER**
> Remember the rules 'change sides, change signs' and 'what you do to one side, you do to the other'.

- To change the subject of a formula, you have to **rearrange** the formula to get the required variable on the left-hand side.
- To rearrange a formula, use the same rules as for solving equations.
- The main difference is that, when rearranging a formula, each step gives an algebraic expression rather than a numerical value.

Make m the subject of $y = 3m - t$.

Add t to both sides: $y + t = 3m$

Reverse the formula: $3m = y + t$

Divide both sides by 3: $m = \dfrac{y + t}{3}$

Questions

Grade A

FM 1 Two families visit the zoo.
The Collins family pay £30 for two adults and three children.
The Gordon family pay £27.50 for one adult and four children.

a How much does it cost for one adult?

b How much does it cost for one child?

Grade C

2 Rearrange the following formulae to make x the subject.

a $T = 4x$ **b** $y = 2x + 3$ **c** $y = \dfrac{x}{5}$

Grade B

3 Rearrange the following formulae to make x the subject.

a $P = 2t + x$ **b** $A = mx + y$ **c** $S = 2\pi x^2$

Algebra

Algebra 2

C

Quadratic expansion

- A **quadratic expression** is one in which the highest power of any term is 2.

  x^2, $4y^2 + 3y$ and $6x^2 - 2x + 1$ are quadratic expressions.

- An expression such as $(x + 2)(x - 3)$ can be **expanded** to give a quadratic expression.

- When an expression such as $(x + 2)(x - 3)$ is multiplied out, it is called **quadratic expansion**.

> **REMEMBER**
> When you multiply out a quadratic expansion, there will always be four terms and two of these terms will combine together.

- There are three methods for quadratic expansion.

  – **Splitting the brackets** The terms inside the first brackets are split and used to multiply the terms in the second brackets.

  Expand $(x + 2)(x + 5)$.
  $(x + 2)(x + 5) = x(x + 5) + 2(x + 5) = x^2 + 5x + 2x + 10 = x^2 + 7x + 10$

  – **FOIL** FOIL stands for First, Outer, Inner and Last. This is the order in which terms are multiplied.

  Expand $(x - 3)(x + 4)$.
  First terms give: $x \times x = x^2$. Outer terms give: $x \times 4 = 4x$.
  Inner terms give: $-3 \times x = -3x$. Last terms give: $-3 \times + 4 = -12$.
  $(x - 3)(x + 4) = x^2 + 4x - 3x - 12 = x^2 + x - 12$

  – **Box method** This is similar to the box method used for long multiplication.

  > **REMEMBER**
  > Be careful with the signs. This is the main place where marks are lost in examination questions involving the expansion of brackets. Remember:
  > $-2 \times -4 = +8$.

  Expand $(x - 4)(x - 2)$.

$\times$	x	$- 4$
x	x^2	$- 4x$
$- 2$	$- 2x$	$+ 8$

  $= x^2 - 4x - 2x + 8$
  $= x^2 - 6x + 8$

Squaring brackets

C

- When you are asked to square a term in brackets, such as $(x + 3)^2$, you must write down the bracket twice, $(x + 3)(x + 3)$, and then use whichever method you prefer to expand the brackets.

  Expand $(x + 3)^2$.
  $(x + 3)^2 = (x + 3)(x + 3) = x^2 + 3x + 3x + 9 = x^2 + 6x + 9$

Questions

Grade C

1 Expand the following expressions.

  a $(x - 1)(x + 3)$ **b** $(m + 2)(m - 6)$

  c $(n + 1)(n - 2)$ **d** $(x + 5)(x + 1)$

  e $(x - 3)(x - 3)$ **f** $(x + 3)(x + 7)$

Grade C

2 Expand the following squares.

  a $(x + 5)^2$ **b** $(x - 2)^2$

Grade C

3 Expand the following expressions.

  a $(x - 1)(x + 1)$ **b** $(m + 2)(m - 2)$

Factorising quadratic expressions

Factorising a quadratic with a unit coefficient of x^2

- **Factorisation** involves putting a quadratic expression back into its **brackets** (if possible).
- When an expression such as $x^2 + ax + b$ is factorised, it is always of the form $(x + p)(x + q)$.

  $x^2 + ax + b = (x + p)(x + q)$, where $p + q = a$ and $pq = b$.

- You can pick up clues about the signs in the brackets from the signs in the quadratic equation.

  $x^2 + ax + b = (x + p)(x + q)$ $x^2 - ax + b = (x - p)(x - q)$ $x^2 \pm ax - b = (x + p)(x - q)$

  > Factorise $x^2 - x - 2$.
  >
  > You know that the factorisation must be of the form $(x + p)(x - q)$, so look for two numbers, p and q, such that $p + q = -1$ and $pq = -2$. So, $x^2 - x - 2 = (x + 1)(x - 2)$.

- **Factorising** a quadratic expression with a coefficient in front of the x^2 term is more complicated.
- When an expression such as $ax^2 + bx + c$ is factorised, it is always of the form $(mx + p)(nx + q)$.

  $ax^2 + bx + c = (mx + p)(nx + q)$, where $mn = a$ and $pq = c$.

  > Factorise $3x^2 + 5x - 2$.
  >
  > You know the factorisation must be of the form $(mx - p)(nx + q)$, so you need to find a combination of m, n, p and q that will give the right number of xs. Considering the factors of 2 (1×2) and 3 (1×3), the solution is $3x^2 + 5x - 2 = (3x - 1)(x + 2)$.

Difference of two squares

- Multiplying out $(x + a)(x - a)$ gives $x^2 - a^2$ because the x terms cancel each other out.
- This expansion is called the **difference of two squares**. It can be used to factorise certain special quadratic expressions.
- Reversing the above, $x^2 - a^2 = (x + a)(x - a)$, where a^2 is a square number.

  > Factorise $x^2 - 9$.
  >
  > Recognise the difference of two squares, x^2 and 3^2. So it factorises to $(x + 3)(x - 3)$.

Solving quadratic equations by factorisation

- Once a quadratic expression has been factorised, it can be used to solve the equivalent quadratic equation.

  > Solve $x^2 + 6x + 8 = 0$.
  >
  > This factorises into $(x + 2)(x + 4) = 0$. Hence, either $x + 2 = 0 \Rightarrow x = -2$ or $x + 4 = 0 \Rightarrow x = -4$. So the solution is $x = -2$ or $x = -4$.

REMEMBER
The equation must be in the form as $ax^2 + bx + c = 0$ before it can be solved.

Questions

Grade B

1 Factorise these expressions.
  a $x^2 + 2x - 3$ **b** $x^2 - 5x + 6$

Grade B

2 Factorise these expressions.
  a $x^2 - 16$ **b** $x^2 - 36$

Grade A

3 Factorise these expressions.
  a $2x^2 - 9x - 5$ **b** $6x^2 + 5x - 6$

Grade B

4 First factorise, then solve these equations.
  a $x^2 - 2x - 3 = 0$ **b** $x^2 + 3x - 10 = 0$

Algebra

Solving quadratic equations

Solving quadratics of the form $ax^2 + bx + c = 0$

- The general quadratic equation is of the form $ax^2 + bx + c = 0$, where a, b and c are positive or negative whole numbers.
- You can solve a general quadratic equation by factorisation, the quadratic formula or completing the square.

Solving a quadratic equation by factorisation

- **Factorise** the quadratic and solve the simple **linear equation** that results when you put each bracket **equal to zero**.

Solve $6x^2 + 5x - 4 = 0$.   This factorises into $(2x - 1)(3x + 4) = 0$.

Hence, either $2x - 1 = 0$ or $3x + 4 = 0 \Rightarrow x = \frac{1}{2}$ or $\Rightarrow x = \frac{-4}{3}$

So the solution is $x = \frac{1}{2}$ or $x = \frac{-4}{3}$.

> **REMEMBER**
> Make sure you can use your calculator to work out the answers.

- When b is zero, the equation can be rearranged to the form $ax^2 = -c$.

Solve $4x^2 - 9 = 0$. $4x^2 - 9 = 0 \Rightarrow 4x^2 = 9 \Rightarrow x^2 = \frac{9}{4} \Rightarrow x = \pm\sqrt{\frac{9}{4}} = \pm\frac{3}{2}$

- When c is zero, the equation is solved by taking out a common factor.

Solve $5x^2 - 15x = 0$. $5x^2 - 15x = 0 \Rightarrow 5x(x - 3) = 0 \Rightarrow x = 0$ or $x = 3$

> **REMEMBER**
> When c is zero, one of the solutions is always $x = 0$.

Solving the general quadratic by the quadratic formula

- Many quadratic equations cannot be solved by factorisation because they do not have simple factors.
- One way to solve this type of equation is to use the **quadratic formula**. This formula can solve any soluble quadratic equation.
- The solutions of the equation $ax^2 + bx + c = 0$ are given by:

$$x = \frac{-b \pm \sqrt{b^2 - 4ac}}{2a}$$

> **REMEMBER**
> The quadratic formula is given on the formula sheet that is included with the examination. However, it is better to learn it because many students miscopy the formula.

Solve $4x^2 + 3x - 2 = 0$.   Take the quadratic formula and put $a = 4$, $b = 3$ and $c = -2$:

$$x = \frac{-(3) \pm \sqrt{(3)^2 - 4(4)(-2)}}{2(4)}$$

Note that the values for a, b and c have been put into the formula in brackets to avoid mistakes in the calculation.

$$x = \frac{-3 \pm \sqrt{9 + 32}}{8} = \frac{-3 \pm \sqrt{41}}{8} \Rightarrow x = 0.43 \text{ or } x = -1.18$$

Questions

Grade A

1 Solve these equations by factorisation.

  a $2x^2 + x - 6 = 0$ b $12x^2 + 17x + 6 = 0$

Grade A

2 Solve these equations. Give your answer in surd form if appropriate.

  a $2x^2 - 7 = 0$ b $4x^2 - 12x = 0$

Grade A

3 Solve $2x^2 - 3x - 1 = 0$, using the quadratic formula. Give your answer to 2 decimal places.

Grade A

AU 4 After correctly substituting numbers into the quadratic formula, Mary gets

$$\frac{5 \pm \sqrt{57}}{4}$$

What quadratic equation is she solving?

Algebra

Completing the square

Solving a quadratic equation by completing the square

- The third method for solving quadratic equations is **completing the square**.

- This method can be used to give answers to a specified number of decimal places or to leave answers in **surd** form.

- The expansion of $(x + a)^2$ is $x^2 + 2ax + a^2$ and rearranging $x^2 + 2ax + a^2 = (x + a)^2$ gives $x^2 + 2ax = (x + a)^2 - a^2$.

- This is the basic principle behind completing the square.

> Rewrite $x^2 + 6x - 5$ in the form $(x + a)^2 - b$.
> We note that: $x^2 + 6x = (x + 3)^2 - 9$
> So, we have: $x^2 + 6x - 5 = (x + 3)^2 - 9 - 5 = (x + 3)^2 - 14$

> **REMEMBER**
> You will notice that the number inside the bracket is half the coefficient of x. In GCSE examinations, the coefficient is always even to avoid problems with fractions.

- Once a quadratic equation has been rewritten by completing the square, it can be used to solve the equation.

> Solve $x^2 + 8x - 5 = 0$.
> First rewrite the left-hand side by the method of completing the square.
> We note that: $x^2 + 8x = (x + 4)^2 - 16$
> So we have: $x^2 + 8x - 5 = (x + 4)^2 - 16 - 5 = (x + 4)^2 - 21$
> So $x^2 + 8x - 5 = 0 \Rightarrow (x + 4)^2 - 21 = 0$
> Rearranging gives: $(x + 4)^2 = 21$
> Taking the square root of both sides gives:
> $x + 4 = \pm \sqrt{21}$
> $x = -4 \pm \sqrt{21}$
> This answer could be left in surd form or evaluated to a given accuracy.

Quadratic equations with no solution

- The quantity $b^2 - 4ac$ in the quadratic formula is known as the **discriminant**.
- When $b^2 - 4ac$ is positive, the equation has two roots or solutions.
- When $b^2 - 4ac$ is zero, the equation has one repeated root or one solution.
- When $b^2 - 4ac$ is negative, the equation has no roots and is not soluble.

> Find the discriminant $b^2 - 4ac$ of the equation $x^2 - 3x + 5 = 0$ and explain what the result tells you.
> $b^2 - 4ac = (-3)^2 - 4(1)(5) = 9 - 20 = -11$, so the equation has no roots.

> **REMEMBER**
> You will not be asked about this in GCSE examinations so, if this happens, you have made a mistake and should check your working.

Questions

Grade A

1 Write an equivalent expression in the form $(x \pm a)^2 - b$.
   a $x^2 + 6x - 1$ **b** $x^2 - 8x - 3$

Grade A*

2 Solve these equations by completing the square. Leave your answers in surd form.
   a $x^2 + 4x - 1 = 0$ **b** $x^2 - 6x + 2 = 0$

Grade A*

AU 3 **a** As part of her coursework, Mary worked out the quadratic equation $x^2 + 6x + 10 = 0$. She could not solve the equation. Explain why.

b The equation Mary should have found was $x^2 + 6x - 10 = 0$. Use the method of completing the square to solve this equation. Leave your answer in surd form.

Algebra

Solving problems with quadratic equations

A

Using the quadratic formula without a calculator

- GCSE examination questions do not specify what method you should use to solve a quadratic equation.

- In the non-calculator paper, you could be asked to solve a quadratic equation that does not factorise. In this case, you would be asked to leave your answer in root or surd form. You could use completing the square, but sometimes the quadratic formula is easier to use.

> Solve $x^2 - x - 11 = 0$.
>
> This equation does not factorise and the coefficient of x is odd so completing the square will give fractional values.
>
> Putting $a = 1$, $b = -1$ and $c = -11$ into the quadratic formula gives:
>
> $$x = \frac{-(-1) \pm \sqrt{(-1)^2 - 4(1)(-11)}}{2(1)} = \frac{1 \pm \sqrt{45}}{2} = \frac{1 \pm 3\sqrt{5}}{2}$$

A*

Solving problems with quadratic equations

- Many real-life problems can be solved by modelling with quadratic equations.

> The right-angled triangle shown has sides of $x + 1$, $4x$ and $5x - 1$.
> Find the value of x.
>
>
>
> As the triangle is right-angled, the sides must obey Pythagoras' theorem.
> So, $(5x - 1)^2 = (4x)^2 + (x + 1)^2$
> $\Rightarrow 25x^2 - 10x + 1 = 16x^2 + x^2 + 2x + 1$
> $\Rightarrow 8x^2 - 12x = 0$
> $\Rightarrow 4x(2x - 3) = 0$
> $\Rightarrow x = 0$ or $x = 1.5$
> x cannot be 0, so $x = 1.5$.

REMEMBER

The quadratic equation you end up with will always factorise. If it does not, go back and check your working.

Questions

Grade A

1 Use the quadratic formula to solve these equations. Leave your answers in surd form.

   a $x^2 - 3x - 1 = 0$ **b** $2x^2 + 6x - 1 = 0$

Grade A*

2 **a** Show that $x - \frac{6}{x} = 1$ can be rearranged to $x^2 - x - 6 = 0$.

   b Solve $x^2 - x - 6 = 0$.

Grade A*

3 The sides of a rectangle are $(3x - 2)$ m and $(2x + 3)$ m. The area of the rectangle is 28 m². What is the value of x?

Grade A*

4 Solve the equation $(x + 2)^2 + (2x - 1)^2 = (3x + 4)^2$.

Algebra

Real-life graphs

Travel graphs

- A **travel graph** shows information about a journey.

- Travel graphs are also known as **distance-time** graphs.

- Travel graphs show the main features of a journey and use **average speeds**, which is why the lines in them are straight.

This graph shows a car journey from Barnsley to Nottingham and back.

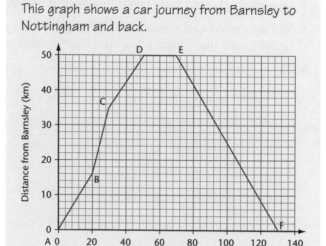

- In **reality**, vehicles do not travel at constant speeds.

- The average speed is given by:

$$\text{average speed} = \frac{\text{total distance travelled}}{\text{total time taken}}$$

- In a travel graph, the **steeper** the line, the **faster** the vehicle is travelling.

REMEMBER

When asked for an average speed, give the answer in kilometres per hour (km/h) or miles per hour (mph).

Real-life graphs

- Some situations can lead to unusual graphs.

This graph shows the depth of water in a conical flask as it is filled from a tap delivering water at a steady rate.

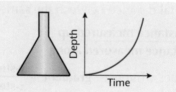

Questions

Grade D

FM 1 Refer to the travel graph above.

  a After how many minutes was the car 16 kilometres from Barnsley?

  b What happened between points D and E?

  c Between which two points was the car travelling fastest?

  d What was the average speed for the part of the journey between C and D?

Grade B

AU 2 Draw a graph of the depth of water in these containers as they are filled steadily.

  a   **b**

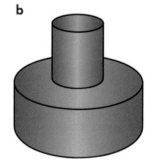

Algebra

Linear graphs

D

Linear graphs

- One method of drawing linear graphs is to **plot points**.
- You only need **two points** to draw a straight-line graph. However, it is better to use **three points**, because the the third point acts as a **check**.
- **Plot** the points you have found and **join** them up to draw the line.

> **REMEMBER**
>
> In examinations you are always given a grid and told what range of x-values to use.

> Draw the graph of $y = 3x - 1$.
>
> Pick a value for x, say $x = 3$, and work out the equivalent y-value.
>
> $y = 3 \times 3 - 1 = 8$.
>
> This gives the coordinates $(3, 8)$.
>
> Repeat this for another value of x, such as $x = 1$.
>
> $y = 3 \times 1 - 1 = 2$ giving the coordinates $(1, 2)$.
>
> Choose a third value of x, such as $x = 0$.
>
> $y = 3 \times 0 - 1 = -1$ giving the coordinates $(0, -1)$.

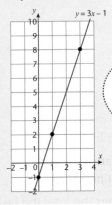

> **REMEMBER**
>
> Always use $x = 0$ as a point as it makes the calculation easy.

C

Gradients

- The **gradient** of a line is a measure of its slope.
- It is calculated by **dividing** the **vertical** distance between two points on the line by the **horizontal** distance between the same two points.

$$\text{gradient} = \frac{\text{distance measured up}}{\text{distance measured along}}$$

- This is sometimes written as: $\text{gradient} = \dfrac{y\text{-step}}{x\text{-step}}$

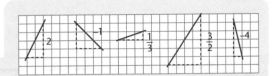

> These lines have gradients as shown.

- Lines that slope down from left to right have **negative** gradients.
- Note the **right-angled triangles** drawn along grid lines. These are used to find gradients.
- To draw a line with a certain **gradient**, for every unit moved **horizontally**, move upwards (or downwards if the gradient is negative) by the number of units of the gradient.

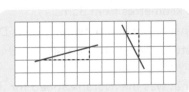

> Draw lines with gradients of $\frac{1}{4}$ and -2.

Questions

Grade D

1 Draw a set of axes with x-values from -3 to $+3$ and y-values from -11 to $+13$. Draw the graph of $y = 4x + 1$ for values of x from -3 to $+3$.

Grade C

2 Find the gradient of the graph you drew in question **1**.

Algebra

The gradient-intercept method

- The **gradient-intercept** method is the most straightforward and quickest method for drawing graphs.

- In the function $y = 2x + 3$, the **coefficient** of x (2) is the **gradient** and the **constant** term (+3) is the **intercept**.

- The **intercept** is the point where the line **crosses the y-axis**.

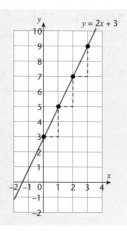

> Draw the graph of $y = 2x + 3$.
>
> First, mark the intercept point $(0, 3)$.
>
> Next, move 1 unit across and 2 units up to show the gradient. Repeat this for each new point.
>
> Finally, join up the points to get the required line.

Cover-up method for drawing graphs

- This method is used to draw graphs of equations in the form $ax + by = c$.
- The x-axis has the equation $y = 0$. This means that all points on the x-axis have a y-value of 0.
- The y-axis has the equation $x = 0$. This means that all points on the y-axis have a x-value of 0.
- These facts can be used to draw any line that has an equation of the form $ax + by = c$.

> Draw the graph of $2x + 3y = 12$.
>
> Because the value of y is 0 on the x-axis, we can solve the equation for x: $2x + 3(0) = 12 \Rightarrow x = 6$
>
> Similarly, because the value of x is 0 on the y-axis, we can also solve the equation for y: $2(0) + 3y = 12 \Rightarrow y = 4$
>
> Hence, the line passes through the points $(6, 0)$ and $(0, 4)$.
>
> Normally we would like a third point, but we can accept two because they are on the axes.
>
>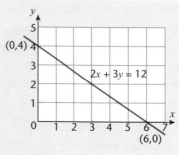
>
> This type of equation can be drawn very easily using the cover-up method.
>
> Start with the equation $2x + 3y = 12$
>
> Cover up the x-term: $(2x) + 3y = 12 \Rightarrow$ when $x = 0$, $y = 4$
>
> Cover up the y-term: $2x + (3y) = 12 \Rightarrow$ when $y = 0$, $x = 6$
>
> This gives the points $(6, 0)$ and $(0, 4)$.

Questions

Grade C

1 Here are the equations of four lines.

　A: $y = 2x - 3$　　　B: $y = 3x - 3$

　C: $y = 2x + 1$　　　D: $y = \frac{1}{2}x - 1$

　a Which two lines are parallel?

　b Which two lines cross the y-axis at the same point?

Grade C

2 a Draw a set of axes with x-values from -3 to $+3$ and y-values from -9 to $+15$. On these axes, draw the graph of $y = 4x + 3$.

b Draw a set of axes with x-values from -4 to $+4$ and y-values from -1 to $+3$. On these axes, draw the graph of $y = \frac{1}{2}x + 1$.

Grade B

3 Draw these lines using the cover-up method. Use the same grid, taking x from 0 to 6 and y from 0 to 6.

　a $3x + 4y = 12$　　　**b** $2x + 5y = 10$

Grade B

AU 4 Which of the following is the odd one out? Give a reason for you answer.

　A: $y = \frac{1}{2}x + 3$　B: $2y - x = 8$　C: $y = 2x + 4$

Algebra

Equations of lines

Finding the equation of a line from its graph

B

- When a graph is expressed in the form $y = mx + c$, the coefficient of x, m, is the **gradient**, and the constant term, c, is the **intercept** on the y-axis.

- This means that if we can find the gradient, m, of a line and its intercept, c, on the y-axis, we can write down the equation of the line.

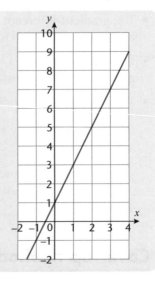

Find the equation of the line shown.

First, find the y-intercept. This is at $(0, 1)$, so $c = 1$.

Next, measure the gradient of the line. For every move of 1 unit across, we move 2 units up which is a gradient of 2, so $m = 2$.

Hence the equation of the line is $y = 2x + 1$.

Uses of graphs – finding formulae or rules

B

- You revised real-life graphs on page 81. You can obtain formulae and equations from these graphs.

> **REMEMBER**
> Be careful with different scales on the axes when working out gradients.

This graph shows the cost of mobile phone calls on a monthly tariff.

The graph shows that there is a fixed charge of £10 per month. This is shown on the graph by the y-intercept (10). There is an additional charge of 20p per minute for all calls. This is shown by the gradient of the line, which is $\frac{1}{5}$.

Hence the equation for the cost of calls (in pounds) is $C = \frac{1}{5}m + 10$, where m is the number of minutes of calls.

So, for 60 minutes of calls the cost is $C = \frac{1}{5} \times 60 + 10 = £22$, which can be checked from the graph.

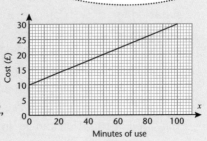

Questions

Grade B

1 What is the equation of the line shown?

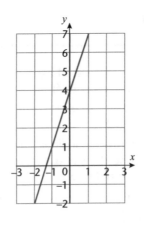

Grade B

FM 2 This graph shows the cost of a taxi journey.

- **a** How much will a journey of 3 km cost?
- **b** What is the gradient of the line?
- **c** What is the intercept with the cost axis?
- **d** Write down an equation for the cost of a journey, C, in terms of the distance, d.
- **e** Work out the value of the equation in part **d** when $d = 3$.

Algebra

Linear graphs and equations

Uses of graphs – solving simultaneous graphs

- You can find the solution to two simultaneous equations by drawing their graphs on the same pair of axes.

By drawing their graphs on the same grid, find the solution to these simultaneous equations. $2x + y = 6$
$$y = 2x$$

First, draw the graph $2x + y = 6$ using the cover-up method.

It crosses the x-axis at $(3, 0)$ and the y-axis at $(0, 6)$.

Next, draw the graph of $y = 2x$ using the gradient-intercept method.

It passes through the origin and has a gradient of 2.

The graphs intersect at $(1.5, 3)$. So the solution to the simultaneous equations is $x = 1.5$, $y = 3$.

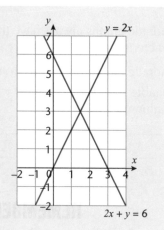

Parallel and perpendicular lines

- If two lines are **parallel** their gradients are the same.

- If two lines are **perpendicular** their gradients are **negative reciprocals** of each other. For example, lines with a gradient of 2 and $-\frac{1}{2}$ are perpendicular to each other.

> **REMEMBER**
> Questions on perpendicular lines are usually accompanied by a diagram.

Find the equation of the line perpendicular to AB that passes through its mid-point

Find the mid-point of AB: M $(2, 3)$

Next, find the gradient of the line AB: 2 and work out the gradient of the perpendicular line: $-\frac{1}{2}$

Now, find the y-intercept. This can be done by drawing the line on the grid or by substituting the known point $(2, 3)$ into $y = mx + c$.

$3 = -\frac{1}{2} \times 2 + c \Rightarrow 3 = -1 + c \Rightarrow c = 4$

Hence the equation of the required line is $y = -\frac{1}{2}x + 4$.

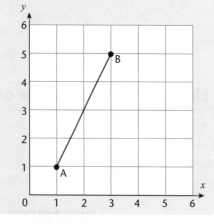

Questions

Grade B

1 By drawing their graphs on the axes supplied, solve these simultaneous equations.

$y = 2x - 1$
$3x + 2y = 12$

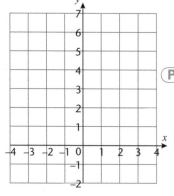

Grade A

2 Find the equation of the line perpendicular to $y = \frac{1}{3}x + 2$ which passes through the point $(0, 5)$.

Grade A*

PS 3 Find the equation of the perpendicular bisector of the points A$(3, 8)$ and B$(7, 6)$.

Algebra

Quadratic graphs

Drawing quadratic graphs

- A quadratic graph has an x^2 term in its equation.

  $y = x^2$ and $y = x^2 + 2x - 3$ will give quadratic graphs.

- Quadratic graphs always have the same characteristic shape, which is called a parabola.

- Quadratic graphs are drawn from tables of values.

  This table shows the values of $y = x^2 + 2x - 3$ for values of x from -4 to 2.

x	-4	-3	-2	-1	0	1	2
y	5	0	-3	-4	-3	0	5

> **REMEMBER**
>
> As all parabolas have line symmetry, there will be some symmetry in the y-values in the table.

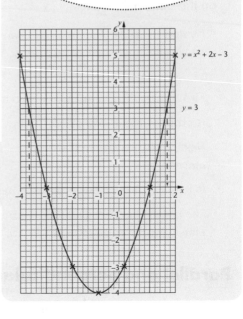

> **REMEMBER**
>
> Try to draw a smooth curve through all the points. Examiners prefer a good attempt at a curve rather than points joined with a ruler.

Reading values from quadratic graphs

- Once a quadratic graph is drawn it can be used to solve various equations.

  Use the graph of $y = x^2 + 2x - 3$ to find the x-values when $y = 3$.

  Draw the line $y = 3$ and draw down to the x-axis from the points where the line intercepts the curve, as shown above.

  The x-values are about $+1.7$ and -3.7.

Using graphs to solve quadratic equations

- Solving a quadratic equation means finding the x-values that make it true.

- To solve a quadratic equation from its graph, read the values where the curve crosses the x-axis.

> **REMEMBER**
>
> The graph is $y = x^2 - 3x - 4$ and the equation is $x^2 - 3x - 4 = 0$. The solution to the equation is found where the graph crosses the x-axis, where $y = 0$.

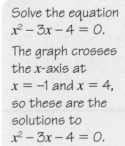

Solve the equation $x^2 - 3x - 4 = 0$.

The graph crosses the x-axis at $x = -1$ and $x = 4$, so these are the solutions to $x^2 - 3x - 4 = 0$.

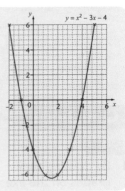

Questions

Grade C

1 **a** Set up and complete the table of values for $y = x^2 + x - 1$ for $-3 \leqslant x \leqslant 3$.

   b Draw the graph of $y = x^2 + x - 1$. Label the x-axis from -3 to $+3$ and the y-axis from -2 to 12.

   c Use the graph to find the values of x when $y = 2$.

   d Use the graph to solve the equation $x^2 + x - 1 = 0$.

The significant points of a quadratic graph

- A quadratic graph has **four points** that are of interest to a mathematician.

  - The points A and B, where the graph crosses the x-axis, are called the **roots**

  - The point C, where the graph crosses the y-axis, is called the **intercept**

  - The point D, which is the maximum or minimum point of the graph, is called the **vertex**

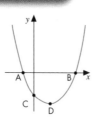

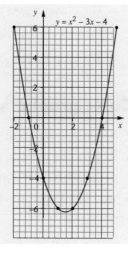

The graph of $y = x^2 - 3x - 4$ is shown.

a Use the graph to find the solutions to $x^2 - 3x - 4 = 0$.
The roots are -1 and 4.

b Where does the graph cross the y-axis? The intercept is -4.

c What are the coordinates of the vertex?
The vertex is at $(1.5, -6.25)$.

Solving equations by the method of intersection

- Many equations can be solved by drawing two intersecting graphs on the same axes and using the x-value(s) of their point(s) of intersection.

- This usually involves the intersection of a curve (usually a quadratic) and a straight line that is used to solve a new equation.

- The method for finding the straight line has four steps.

Given the graph of $y = x^2 + 3x - 2$, draw a suitable straight line to solve the equation $x^2 + 2x - 3 = 0$.
- Step 1: Write down the original (given) equation
- Step 2: Write down the (new) equation to be solved in reverse
- Step 3: Subtract these equations
- Step 4: Draw the line given by the answer

$$y = x^2 + 3x - 2$$
$$- \ 0 = x^2 + 2x - 3$$
$$\overline{\quad y = x + 1 \quad}$$

Questions

Grade A–A*

1 The graph of $y = x^2 + 3x - 2$ is shown opposite.

  a Write down the coordinates of the intercept with the y-axis.

  b Write down the coordinates of the vertex.

  c What are the roots of the equation $x^2 + 3x - 2 = 0$?

  d By drawing a suitable straight line, solve the equation $x^2 + 3x - 1 = 0$.

  e By drawing another straight line, solve the equation $x^2 + 2x - 2 = 0$.

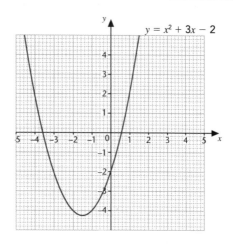

Algebra

Reciprocal graphs

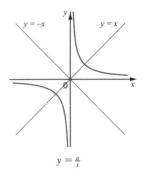

- A reciprocal equation has the form $y = \frac{a}{x}$.
- All reciprocal graphs have a similar shape and some symmetry properties.
  - The lines $y = x$ and $y = -x$ are lines of symmetry
  - The closer x gets to zero, the nearer the graph gets to the y-axis
  - As x increases, the graph gets closer to the x-axis
  - The graph never actually touches the axes, but gets closer and closer to them: the axes are called **asymptotes**

Cubic graphs

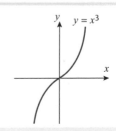

- A cubic function or graph is one that contains a term in x^3.
- You should be able to recognise the characteristic shape of the basic function $y = x^3$.
- You should be able to draw other cubic graphs from a table of values.

This partly completed table shows $y = x^3 - x$ for values of x from -3 to 3.

x	-3	-2.5	-2	-1.5	-1	-0.5	0	0.5	1	1.5	2	2.5	3
y	-24	-13.13	-6	-1.88				-0.38	0	1.88	6		

Exponential graphs

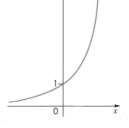

- Equations that have the form $y = k^x$, where k is a positive number, are called **exponential functions**.
- Exponential functions share the following properties.
  - When k is greater than 1, the value of y increases steeply as x increases
  - Also, when k is greater than 1, as x takes on increasing large negative values, the closer y gets to zero, and so the negative x-axis becomes an asymptote
  - Whatever the value of k, the graph always crosses the y-axis at 1
  - The reciprocal graph, $y = k^{-x}$, is the reflection in the y-axis of $y = k^x$
  - When k is less than 1, the graph is the same shape as $y = k^{-x}$
- Exponential graphs are always drawn from a table of values.

This partly completed table shows $y = 2^x$ for values of x from 0 to 4.

x	0	0.5	1	1.5	2	2.5	3	3.5	4
y	1	1.41	2	2.83	4				16

Questions

Grade A

1 Complete the table of values for $y = x^3 - x$ above and use it to draw the graph of $y = x^3 - x$ for values of x from -3 to $+3$. (Take the y-axis from -24 to $+24$.)

2 Complete the table for $y = 2^x$ above and use it to draw the graph of $y = 2^x$ for values of x from 0 to 4. (Take the y-axis from 0 to $+16$.)

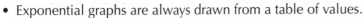

Algebraic fractions

Algebraic fractions

- The following four rules are used to work out the value of fractions.

Addition: $\quad \frac{a}{b} + \frac{c}{d} \equiv \frac{ad + bc}{bd}$

Subtraction: $\quad \frac{a}{b} - \frac{c}{d} \equiv \frac{ad - bc}{bd}$

Multiplication: $\quad \frac{a}{b} \times \frac{c}{d} \equiv \frac{ac}{bd}$

Division: $\quad \frac{a}{b} \div \frac{c}{d} \equiv \frac{ad}{bc}$

- Note that a, b, c and d can be numbers, letters or algebraic expressions.
- The following points make working with expressions easier.
  - Use **brackets**
  - **Factorise** if you can
  - **Cancel** if you can

> **REMEMBER**
> Use brackets around terms in the rules before you work them out. This makes it less likely that you will make a mistake.

Simplify $\dfrac{2y}{x} + \dfrac{x}{2y}$.

Using the addition rule: $\quad \dfrac{(2y)(2y) + (x)(x)}{(x)(2y)} = \dfrac{4y^2 + x^2}{2xy}$

Simplify $\dfrac{2}{x} - \dfrac{y}{3x}$.

Using the subtraction rule: $\quad \dfrac{(2)(3x) - (x)(y)}{(x)(3x)} = \dfrac{6x - xy}{3x^2} = \dfrac{x(6 - y)}{3x^2} = \dfrac{6 - y}{3x}$

Simplify $\dfrac{a}{2} \times \dfrac{a + b}{a - b}$.

Using the multiplication rule: $\quad \dfrac{(a)(a + b)}{(2)(a - b)} = \dfrac{a^2 + ab}{2a - 2b}$

> **REMEMBER**
> You will be expected to simplify answers as much as possible, so remember to look out for common factors.

Simplify $\dfrac{a}{4} \div \dfrac{3a}{8}$.

Using the division rule: $\quad \dfrac{(a)(8)^2}{(4)^1(3a)} = \dfrac{2}{3}$

Questions

Grade A

1 Simplify each of these.

  a $\dfrac{x + 1}{2} + \dfrac{x - 2}{3}$ **b** $\dfrac{x}{5} - \dfrac{2x + 1}{3}$

  c $\dfrac{2x}{5} \times \dfrac{3y}{4}$ **d** $\dfrac{4x}{3y} \div \dfrac{x}{y}$

Grade A

2 Simplify each of these.

  a $\dfrac{x}{5} - \dfrac{x - 1}{7}$ **b** $\dfrac{4x^2}{6y} \times \dfrac{2y^3}{5x}$ **c** $\dfrac{x - 2}{3} \times \dfrac{9}{3x - 6}$

Grade A*

3 Simplify each of these.

  a $\dfrac{5}{x + 2} - \dfrac{2}{x - 3}$ **b** $\dfrac{x}{x - 2} - \dfrac{3}{x + 4}$

A*

Algebra

Solving equations

Solving equations with algebraic fractions

A*

- The algebraic fraction identities on page 89 can be used to solve equations.

Solve $\dfrac{x+1}{3} - \dfrac{x-3}{2} = 1$.

Use the subtraction rule for combining fractions, and also cross-multiply the denominator of the left-hand side to the right-hand side.

$\dfrac{(x+1)(2) - (3)(x-3)}{(3)(2)} = 1 \Rightarrow 2(x+1) - 3(x-3) = (1)(3)(2)$

$2x + 2 - 3x + 9 = 6 \Rightarrow -x + 11 = 6 \Rightarrow -x = -5 \Rightarrow x = 5$

> **REMEMBER**
> Use brackets around all numbers and expressions. Most marks are lost in this type of problem because of errors with minus signs.

Solve $\dfrac{3}{x-1} - \dfrac{2}{x+1} = 1$.

Use the subtraction rule for combining fractions, and cross-multiply the denominator, using the method shown in the previous example.

$(3)(x+1) - (2)(x-1) = (x-1)(x+1)$

$3x + 3 - 2x + 2 = x^2 - 1$

(Right-hand side is the difference of two squares.)

Rearrange into the general quadratic form.

$x^2 - x - 6 = 0$

Factorise and solve the equation.

$(x-3)(x+2) = 0 \Rightarrow x = 3 \text{ or } x = -2$

> **REMEMBER**
> When you rearrange your equation into the quadratic it should factorise. If it doesn't, then you will almost certainly have made a mistake unless the question states that it requires an answer as a surd or a decimal. Check your minus signs

Simplify $\dfrac{2x^2 + x - 3}{4x^2 - 9}$.

Factorise the numerator and denominator:

$\dfrac{(2x+3)(x-1)}{(2x+3)(2x-3)}$

Cancel any common factors:

$\dfrac{x-1}{2x-3}$

> **REMEMBER**
> There will always be a common factor on the top and the bottom which will cancel out.

> **REMEMBER**
> Watch out for the difference of two squares. This is a favourite with examiners in this sort of question! For example $4x^2 - 9 = (2x)^2 - (3)^2$.

Questions

Grade A*

1 Solve these equations.

a $\dfrac{x+1}{2} + \dfrac{x-2}{3} = 4$ **b** $\dfrac{4}{x+1} + \dfrac{5}{x+2} = 2$

Grade A*

2 Simplify these expressions.

a $\dfrac{2x^2 + x - 1}{4x^2 - 1}$ **b** $\dfrac{3x^2 + 2x - 1}{2x^2 + 5x + 3}$

Grade A*

PS 3 Roz has correctly simplified an algebraic fraction for homework.

Unfortunately she has spilt coffee on the page.

$\dfrac{2x^2 - 5x - 3}{x^2}$ $= \dfrac{}{x+1}$

Work out the missing expressions.

Simultaneous equations 2

Linear and non-linear simultaneous equations

- On page 76, you revised solving two linear simultaneous equations by the method of substitution.

- You can use a similar method to solve a pair of a equations, one of which is linear and the other of which is non-linear, but you must always substitute the linear equation into the non-linear equation. This usually gives a soluble quadratic equation.

Solve these simultaneous equations.

$$x^2 + y^2 = 5 \ (1)$$
$$x + y = 3 \ \ (2)$$

Rearrange equation (2) to obtain:

$$y = 3 - x$$

Substitute this into the non-linear equation (1):

$$x^2 + (3 - x)^2 = 5$$

Expand: $\qquad\qquad x^2 + 9 - 6x + x^2 = 5$

Rearrange into the general quadratic form:

$$2x^2 - 6x + 4 = 0$$

Cancel by 2: $\qquad\qquad x^2 - 3x + 2 = 0$

Factorise and solve: $\qquad (x - 1)(x - 2) = 0 \Rightarrow x = 1 \text{ or } x = 2$

Substitute x into equation (2) to find values of y:

when $x = 1, y = 2$

and when $x = 2, y = 1$

So the solutions are (1, 2) and (2, 1).

> **REMEMBER**
> Remember to give your answers as pairs of values in x and y. Marks are often lost in examinations by not giving a full solution.

Solve these simultaneous equations.

$$y = x^2 - 2x \ (1)$$
$$y = 2x - 3 \ (2)$$

In this case, the equations can just be put equal to each other so:

$$x^2 - 2x = 2x - 3$$

Rearrange into the general quadratic form:

$$x^2 - 4x + 3 = 0$$

Factorise and solve: $\qquad (x - 1)(x - 3) = 0 \Rightarrow x = 1 \text{ or } x = 3$

Substitute x into equation (2) to find values of y:

when $x = 1, y = -1$ and when $x = 3, y = 3$

So the solutions are (1, −1) and (3, 3).

> **REMEMBER**
> When you rearrange into a quadratic, it will factorise. If it doesn't you will have made a mistake.

Questions

Grade A*

1 Solve these pairs of simultaneous equations.

a $x^2 + y^2 = 25$
  $x - y = 1$

b $x^2 + y^2 = 13$
  $x = 13 - 5y$

c $y = 2x^2 - x - 9$
  $y = 2x + 5$

d $y = 6x^2 + 9x - 5$
  $y + 7 = 2x$

Algebra

The nth term

nth term of a sequence

- A linear sequence has the same difference between consecutive terms.

  > 3, 8, 13, 18, 23, 28, ... is a linear sequence with a constant difference of 5.

- The nth term of a linear sequence is always of the form $An \pm b$.

  > $3n + 1$, $4n - 3$ and $8n + 7$ are examples of nth terms of a linear sequence.

- To find the coefficient of n, A, find the difference between consecutive terms.

  > The sequence 4, 7, 10, 13, 16, ... has a constant difference of 3, so the nth term of the sequence will be given by $3n \pm b$.

- To find the value of b, work out the difference between A and the first term of the sequence.

  > The sequence 4, 7, 10, 13, 16, ... has a first term of 4. The coefficient of n is 3.
  > To get from 3 to 4 you add 1, so the nth term is $3n + 1$.

  > Find the nth term of the sequence 4, 9, 14, 19, 24, ...
  > The constant difference is 5, and $5 - 1 = 4$, so the nth term is $5n - 1$.

Special sequences

- There are many **special sequences** that you should be able to recognise.
  - The even numbers   2, 4, 6, 8, 10, 12, ... The nth term is $2n$.
  - The odd numbers   1, 3, 5, 7, 9, 11, ... The nth term is $2n - 1$.
  - The square numbers   1, 4, 9, 16, 25, 36, ... The nth term is n^2.
  - The triangular numbers 1, 3, 6, 10, 15, 21, ... The nth term is $\frac{1}{2}n(n + 1)$.
  - The powers of 2   2, 4, 8, 16, 32, 64, ... The nth term is 2^n.
  - The powers of 10   10, 100, 1000, 10 000, 100 000, ... The nth term is 10^n.
  - The prime numbers   2, 3, 5, 7, 11, 13, 17, 19, ... There is no nth term as there is no pattern to the prime numbers.

> **REMEMBER**
> The only even prime number is 2.

Questions

Grade C

1 a The nth term of a sequence is $5n - 1$.

   i Write down the first three terms of the sequence.

   ii Write down the 100th term of the sequence.

  b The nth term of a sequence is $\frac{1}{2}(n + 1)(n + 2)$.

   i Write down the first three terms of the sequence.

   ii Write down the 199th term of the sequence.

Grade C

2 Write down the nth term of each of these sequences.

  a 6, 11, 16, 21, 26, 31, ...

  b 3, 11, 19, 27, 35, ...

  c 9, 12, 15, 18, 21, 24, ...

Grade C

3 a What is the 100th even number?

  b What is the 100th odd number?

  c What is the 100th square number?

  d Continue the sequence of triangular numbers up to 10 terms.

  e Continue the sequence of powers of 2 up to 10 terms.

  f Write down the next five prime numbers after 19.

Grade C

AU 4 Here are two sequences 2, 7, 12, 17, 22, ...
                                      4, 14, 24, 34, 45, ...

Explain why the sequences will never have a term in common.

Algebra

Formulae

The nth term from given patterns

- An important part of mathematics is to find **patterns in situations**.
- Once a **pattern** has been found, the **nth term** can be used to describe the pattern. The diagram shows a series of 'L' shapes.

REMEMBER
Write down the sequence then look for the nth term.

> How many squares are there in L100?
>
> First write down the sequence of squares in the 'L' shapes: 3, 5, 7, 9, …
>
> The constant difference is 2, and 2 + 1 = 3, so the nth term is $2n + 1$.
>
> The 100th term will be $2 \times 100 + 1 = 201$, so there will be 201 squares in L100.

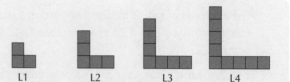

L1   L2   L3   L4

Identities

- You should be able to tell the difference between **expressions**, such as $3x + 7$, **equations**, such as $3x + 7 = 9$, and **formulae**, such as $S = 2n + 1$.
- You should also be able to recognise an **identity**, which is an equality that is **true** for all values of the variable, whether **numerical** or **algebraic**. It is shown using the $\equiv$ symbol.
- The difference of two squares $x^2 - 9 \equiv (x - 3)(x + 3)$ is an identity.

Changing the subject of a formula

- On page 77, you revised changing the **subject** of a formula in which the subject appears once.
- The principle is the same when the the subject appears more than once.

> Make x the subject of $2x + y = ax + b$.
>
> First, rearrange the formula to get all the x terms on the left-hand side and the other terms on the right-hand side: $2x - ax = b - y$
>
> Take a factor of x out of the left-hand side to get: $x(2 - a) = b - y$
>
> Divide by the bracket, which gives: $x = \dfrac{b - y}{2 - a}$

REMEMBER
When the subject appears twice, you will need to factorise and divide by the bracket at some point in your working.

> Make x the subject of $y = \dfrac{x - 2}{x + 3}$
>
> First, cross-multiply which gives: $y(x + 3) = x - 2$
>
> Expand the bracket to get: $yx + 3y = x - 2$
>
> Collect subject terms: $yx - x = -2 - 3y$
>
> Take a factor of x out of the left-hand side to get: $x(y - 1) = -2 - 3y$
>
> Divide by the bracket, which gives: $x = \dfrac{-2 - 3y}{y - 1}$

Questions

Grade C

1 Matches are used to make pentagonal patterns.

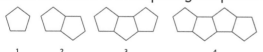

1   2   3   4

a How many matches will be needed to make the 10th pattern?

b How many matches will be needed to make the nth pattern?

Grade A

2 Make x the subject of each of these formulae.

a $ax + 6y = 2x + 8y$ **b** $y = \frac{x + 4}{x - 2}$

Grade B

AU 3 Verify that $x^2 - 9 \equiv (x - 3)(x + 3)$ is an identity by substituting and expanding $x = n + 2$ in both sides.

Algebra

Inequalities

Solving inequalities

- An **inequality** is an algebraic expression that uses the signs $<$ (less than), $>$ (greater than), $\leq$ (less than or equal to) and $\geq$ (greater than or equal to).

- The solutions to inequalities are a **range of values**.

  The expression $x < 2$ means that x can take any value less than 2, all the way to minus infinity. x can also be very close to 2, for example 1.9999..., but is never actually 2.

  $x \geq 3$ means x can take any value greater than 3, up to infinity, or 3 itself.

- **Linear inequalities** can be solved using the same rules that you use to solve equations.

- The answer when a linear inequality is solved is an inequality such as $x > -1$.

  Solve $\frac{x}{5} - 3 \geq 4$.

  Add 3 to both sides: $\frac{x}{5} \geq 7$

  Multiply both sides by 5: $x \geq 35$

> **REMEMBER**
> Don't use the equals sign when solving inequalities as this doesn't get any marks in an examination.

Inequalities on number lines

- The solution to a linear inequality can be shown on a **number line** by using the convention that an open circle is a **strict inequality** and a filled-in circle is an **inclusive inequality**.

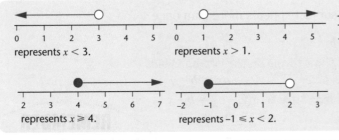

represents $x < 3$. represents $x > 1$. represents $x \leq -2$.

represents $x \geq 4$. represents $-1 \leq x < 2$.

> **REMEMBER**
> Inequalities often ask for integer values. An integer is a positive or negative whole number, including zero.

Solve the inequality $2x + 3 < 11$ and show the solution on a number line.

The solution is $x < 4$, which is shown on this number line.

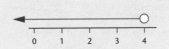

Questions

Grade C

1 Solve the following inequalities.

  a $x + 5 < 8$ **b** $2x + 3 > 5$

  c $\frac{x}{3} - 5 \geq 1$ **d** $\frac{x}{2} + 7 > 2$

Grade C

2 What inequalities are shown by the following number lines?

  a
  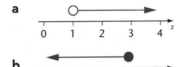

  b

Grade C

3 **a** What inequality is shown on this number line?

  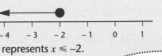

  b Solve the inequality $3x + 6 \geq 3$.

  c What integers satisfy both of the inequalities in parts **a** and **b**?

Grade C

4 Write down the inequality that represents the values of x that satisfy both of the following inequalities.

  $-3 < x \leq 5$ and $-1 \leq x < 8$

Algebra

Graphical inequalities

B

Graphical inequalities

- A linear **inequality** can be plotted on a graph.

- The result is a **region** that lies on one side or the other of a straight **boundary line**.

  > $y < 2, x > -3, y \geq 2x + 1$ and $2x + 3y < 6$ are examples of linear inequalities which can be represented on a graph.

  > **REMEMBER**
  > Use the origin as the point to test if possible as it makes working out the inequality easier.

- The first step is to draw the boundary line. This is found by replacing the inequality sign with an equals sign.

  - If the inequality is strict ($<$ or $>$), the boundary line should be dashed

  - If the inequality is inclusive ($\leq$ or $\geq$), the boundary line should be solid

- Once the boundary line is drawn, the required region is shaded.

  - To confirm on which side of the line the region lies, choose any point that is not on the boundary line and test it in the inequality. If it satisfies the inequality, that is the side required. If it doesn't, the other side is required.

  > Show both of these inequalities on a graph.
  >
  > **a** $y \leq 3$ **b** $y < 2x - 1$
  >
  > First draw the boundary lines:
  >
  > **a** $y = 3$ (solid)   **b** $y = 2x - 1$ (dashed)
  >
  > Test a point that is not on the line, such as $(0, 0)$.

  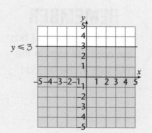
  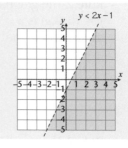

More than one inequality

- You will be required to show more than one inequality on the same graph.

- In this case, it is clearer to shade the regions that are *not* required, so that the required region is left unshaded.

  > On the same grid, show the region bounded by these inequalities.
  >
  > $x \geq 1$ $x + y < 5$ $y > x$
  >
  > Start by drawing the boundary lines:
  >
  > $x = 1$ (solid)
  > $x + y = 5$ (dashed)
  > $y = x$ (dashed)
  >
  > Test a point that is not on the line in each case and shade the appropriate regions.

  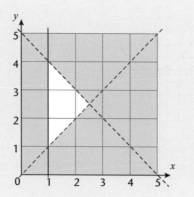

Questions

1 On separate grids, with x-and y-values from 0 to 5, show the following inequalities.

  a $x < 4$ **b** $x + y \leq 2$ **c** $2y + 5x \leq 10$

2 On the same grid with x-and y-values from 0 to 5, show the region bounded by these inequalities.

  $y \leq 3$ $x + y \geq 3$ $y \geq x$

Algebra

Graph transforms

AQA 3   EDEXCEL 3   OCR 3

Transformations of the graph $y = f(x)$

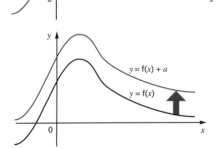

- The notation $f(x)$ is used to represent a **function** of x. A function of x is any algebraic expression in which x is the only variable.

  $f(x) = x + 5$, $f(x) = 7x$, $f(x) = 3x + 2$ and $f(x) = x^2$ are all functions of x.

- The graph $y = f(x)$ can be transformed by six rules that you should understand. Three are covered in this section and three are covered in the next section.

  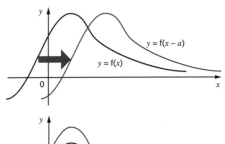

  - **Rule 1** The graph of $y = f(x) + a$ is a translation of the graph $y = f(x)$ by the vector $\binom{0}{a}$

  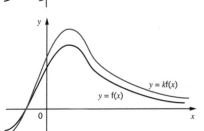

  - **Rule 2** The graph of $y = f(x - a)$ is a translation of the graph $y = f(x)$ by the vector $\binom{a}{0}$

  > **REMEMBER**
  > The vector for $f(x - a)$ has the opposite sign to the number in the bracket.

  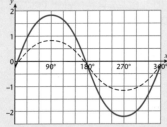

  - **Rule 3** The graph of $y = kf(x)$ is a stretch of the graph $y = f(x)$ by a scale factor of k in the y-direction

The graph of $y = \sin x$ is shown by a dotted line on each of the grids below for $0° \leqslant x \leqslant 360°$. What are the equations of the graphs shown by the solid red lines?

a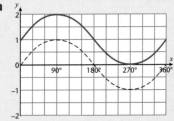

b

a is a translation upwards by 1 unit so the graph is $y = \sin x + 1$.
b is a stretch of scale factor 2 in the y-direction so the graph is $y = 2 \sin x$.

Questions

Grade A*

1 Draw a sketch of $y = \cos x$ for $0° \leqslant x \leqslant 360°$.
On the same axes, sketch the graph of
$y = \cos x - 1$.

Grade A*

2 Draw a sketch of $y = \cos x$ for $0° \leqslant x \leqslant 360°$.
On the same axes, sketch the graph of
$y = \frac{1}{2} \cos x$.

Grade A*

3 Draw a sketch of $y = x^2$. Label both axes
-5 to $+5$.

On the same axes, sketch the graphs

a $y = (x - 1)^2$

b $y = x^2 + 1$

c $y = 2x^2$

Algebra

Transformations of the graph $y = f(x)$

- The graph $y = f(x)$ can be transformed by six rules that you should understand. Rules 4 to 6 are covered here.

  - **Rule 4** The graph of $y = f(tx)$ is a stretch of the graph $y = f(x)$ by a scale factor of $\frac{1}{t}$ in the x-direction.

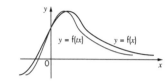

  - **Rule 5** The graph of $y = -f(x)$ is a reflection of the graph $y = f(x)$ in the x-axis.

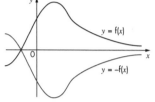

> **REMEMBER**
>
> The scale factor of the stretch is the reciprocal of the number inside the bracket.

  - **Rule 6** The graph of $y = f(-x)$ is a reflection of the graph $y = f(x)$ in the y-axis.

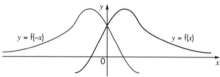

The graph of $y = \sin x$ is shown by a dotted line on each of the grids below for $0° \leqslant x \leqslant 360°$.

What are the equations of the graphs shown by the solid red lines?

a
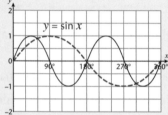

b
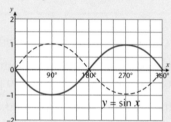

a is a stretch by scale factor $\frac{1}{2}$ in the x-direction, so the graph is $y = \sin 2x$.

b is a reflection in the x-axis, so the graph is $y = -\sin x$.

The graph of $y = x^3 + 1$ is shown by a dotted line on the axes on the right.

On the same axes, sketch the graph of $y = -x^3 + 1$.

The graph is $y = (-x)^3 + 1$ which is a reflection of the curve in the y-axis. This is the graph shown by a solid line.

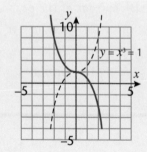

Questions

Grade A*

1 Draw a sketch of $y = \cos x$ for $0° \leqslant x \leqslant 360°$. On the same axes, sketch the graph of $y = \cos 3x$.

Grade A*

2 Draw a sketch of $y = \cos x$ for $0° \leqslant x \leqslant 360°$. On the same axes, sketch the graph of $y = -\cos x$.

Grade A*

3 Draw a sketch of $y = x^3$. Label the x-axis from -5 to $+5$ and the y-axis from -10 to $+10$. On the same axes sketch

  a $y = x^3 + 1$

  b $y = \frac{1}{2}x^3$

  c $y = (x + 2)^3$

Algebra

Proof

Proof

- The method of mathematical **proof** is to proceed in logical steps, establishing a series of mathematical statements by using facts that are already known to be true.

- Each step must follow from the previous step and all statements must be given reasons if necessary.

Geometric proof

- Geometric proofs are usually concerned with circle theorems or congruency.

- They use the four conditions for congruency – SSS, SAS, ASA and RHS – which you revised on page 60.

- In a geometric proof you must give reasons for the values you find for the angles.

ABCD is a parallelogram. X is the point where the diagonals meet. Prove that triangles AXB and CXD are congruent.

$\angle BAX = \angle DCX$ (alternate angles)

$\angle ABX = \angle CDX$ (alternate angles)

$AB = CD$ (opposite sides in a parallelogram)

Hence $\triangle AXB$ is congruent to $\triangle CXD$ (ASA).

Algebraic proof

- Algebraic proofs are usually concerned with proving that two algebraic expressions are equivalent.

- You usually start with the left-hand expressions and use the standard algebraic techniques of expansion, simplification, factorisation and cancelling to prove that this is equal to the right-hand side.

- You must make sure that all the algebra is well explained and each step is clear. Do not assume that an examiner will fill in the gaps for you.

Prove that $(n + 1)^2 + n^2 - (n - 1)^2 = n(n + 4)$.

Expand the LHS to get: $\qquad n^2 + 2n + 1 + n^2 - (n^2 - 2n + 1)$

Show clearly that the minus sign in front of the second bracket is properly dealt with:

$$n^2 + 2n + 1 + n^2 - n^2 + 2n - 1$$

Collect like terms: $\qquad n^2 + 4n$

Factorise the collected result: $n^2 + 4n = n(n + 4)$ which is the RHS of the original expression.

Questions

PS 1 Prove that the exterior angle of a triangle, z, is equal to the sum of the two opposite angles, x and y.

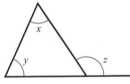

PS 2 Prove that $(n + 6)^2 - (n + 2)^2 = 8(n + 4)$.

PS 3 ABC is an isosceles triangle. M is the mid-point of the base AB.

Prove that ACM and BCM are congruent triangles.

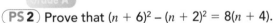

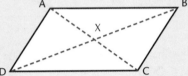

Algebra grade booster

I can...

- [ ] recognise expressions, equations and formulae
- [ ] use letters to write more complicated algebraic expressions
- [ ] expand expressions with brackets
- [ ] factorise simple expressions
- [ ] solve linear equations where the variable appears on both sides of the equals sign
- [ ] solve linear equations that require the expansion of a bracket
- [ ] set up and solve simple equations from real-life situations
- [ ] find the average speed from a travel graph
- [ ] draw a linear graph without a table of values
- [ ] substitute numbers into an nth-term rule
- [ ] understand how odd and even numbers interact in addition, subtraction and multiplication problems

You are working at **Grade D** level.

- [ ] expand and simplify expressions involving brackets
- [ ] factorise expressions involving letters and numbers
- [ ] expand pairs of linear brackets to give a quadratic expression
- [ ] solve linear equations that have the variable on both sides and include brackets
- [ ] solve simple linear inequalities
- [ ] show inequalities on a number line
- [ ] solve equations, using trial and improvement
- [ ] rearrange simple formulae
- [ ] give the nth term of a linear sequence
- [ ] give the nth term of a sequence of powers of 2 or 10
- [ ] draw a quadratic graph, using a table of values

You are working at **Grade C** level.

- [ ] solve a quadratic equation from a graph
- [ ] recognise the shape of the graph $y = x^3$
- [ ] solve two linear simultaneous equations
- [ ] rearrange more complex formula to make another variable the subject
- [ ] factorise a quadratic expression of the form $x^2 + ax + b$
- [ ] solve a quadratic equation of the form $x^2 + ax + b = 0$
- [ ] interpret real-life graphs
- [ ] find the equation of a given linear graph
- [ ] solve a pair of linear simultaneous equations from their graphs
- [ ] draw cubic graphs, using a table of values
- [ ] use the nth term to generate a quadratic sequence

Algebra

☐ solve equations involving algebraic fractions where the subject appears as the numerator

☐ solve more complex linear inequalities

☐ represent a graphical inequality on a coordinate grid

☐ find the inequality represented by a graphical inequality

☐ verify results by substituting numbers

You are working at (Grade B) level.

☐ recognise the shape of the graph $y = \frac{1}{x}$

☐ draw exponential and reciprocal graphs, using a table of values

☐ find the proportionality equation from a direct or inverse proportion problem

☐ set up and solve two simultaneous equations from a practical problem

☐ factorise a quadratic expression of the form $ax^2 + bx + c$

☐ solve a quadratic equation of the form $ax^2 + bx + c = 0$ by factorisation

☐ solve a quadratic equation of the form $ax^2 + bx + c = 0$ by the quadratic formula

☐ write a quadratic expression of the form $x^2 + ax + b$ in the form $(x + p)^2 - q$

☐ interpret and draw more complex real-life graphs

☐ find the equations of graphs parallel and perpendicular to other lines and passing through specific points

☐ rearrange a formula where the subject appears twice

☐ combine algebraic fractions, using the four rules of addition, subtraction, multiplication and division

☐ translate and solve a real-life problem, using inequalities

☐ show that a statement is true, using verbal or mathematical arguments

You are working at (Grade A) level.

☐ solve equations using the intersection of two graphs

☐ use trigonometric graphs to solve sine and cosine problems

☐ solve direct and inverse proportion problems, using three variables

☐ solve a quadratic equation of the form $x^2 + ax + b = 0$ by completing the square

☐ solve real-life problems that lead to a quadratic equation

☐ solve quadratic equations involving algebraic fractions where the subject appears as the denominator

☐ rearrange more complicated formula where the subject may appear twice

☐ simplify algebraic fractions by factorisation and cancellation

☐ solve a pair of simultaneous equations where one is linear and one is non-linear

☐ transform the graph of a given function

☐ identify the equation of a transformed graph

☐ prove algebraic results with rigorous and logical arguments

You are working at (Grade A*) level.

Algebra

Answers

Page 16

1 a mode = 8, median = 7.5, mean = 6.6

  b mode = 11, median = 12, mean = 12

2 4, 4, 5, 6, 11 (other answers possible)

3 a 30   b 0   c 1   d i 45 ii 1.5

Page 17

1 a $20 < x \le 30$

  b i 1350 ii 27

2 a 10

  b Because there are 25 snails below 15 and 75 above 15

  c 27.5 grams

3 a Could be true. The greatest possible range is 60 − 5 = 55 g and the smallest possible range is 55 − 10 = 45 g.

  b False. There are 15 between 30 and 40 and 30 between 40 and 60. The bars are the same height but the frequency is represented by the area of the bars not the height.

  c True. $10 \times 7.5 + 15 \times 12.5 + 30 \times 22.5 + 45 \times 45 = 2962.5$, $2962.5 \div 100 \approx 29.6$g.

Page 18

1 For the outside courgettes the mean is 13.5 and the range is 9.

The mean of the greenhouse courgettes is 15.5 so 2 cm longer.

The range of the greenhouse courgettes is 6 so 3 cm less.

The data supports the hypothesis that the courgettes grown in the greenhouse are bigger, although the sample size is small. The greenhouse courgettes are also more consistent than the outside courgettes as the range is smaller.

2 The headteacher will have to collect data from previous years' GCSE results and compare the numbers of grades for boys and girls.

Using results from last year, the headteacher could allocate a points score of 8 for an A* and 1 for a G for all maths GCSEs obtained by boys and girls.

The headteacher could then calculate a mean score for boys and girls. The number of each grade obtained by boys and girls could be shown on a dual bar chart.

If the mean score for the girls is higher than the boys, then the hypothesis is true, but if the values are quite close or the boys have a higher mean than the girls, the hypothesis is not true.

Page 19

1 a One of: leading question, two questions in one, double negative, not enough responses

  b One of: overlapping responses, missing responses

2 a i Not representative as only year 7

    ii Could be random but students who walk are excluded

    iii This will give a random sample as all students have an equal chance.

  b Y7: 12, Y8: 10, Y9: 12, Y10: 11, Y11: 9

Page 20

1 a Town B – on average about 4 degrees hotter

  b Town B – December is in summer *or* June is in winter

  c September

  d No – the lines cross but they have no meaning (just show trends)

2 a 20

  b 36 (49 – 13)

  c 35

  d 26.5

  e 28 (560 ÷ 20)

Page 21

1 a positive

  b The higher the spelling marks, the higher the tables marks

2 a As the car gets older, its value decreases

  b There is no relationship between the distance someone lives from work and the wages they earn

3 This will be a follow through from your line. Value should be 70 – 75.

Page 22

1 a 66

  b 48

  c 83

  d 35

  e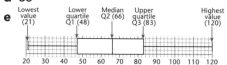

2 The box plot will be symmetrical about the median.

Page 23

1 a Ali: 0.25 Barry: 0.22 Clarrie: 0.19
   b Clarrie – most trials
   c 40
2 a i $\frac{1}{2}$ ii $\frac{1}{2}$
   iii They cannot happen at same time
   iv The probabilities add up to 1
   b i $\frac{1}{13}$ ii $\frac{12}{13}$

Page 24

1 a 50   b 20
   c 70
2 a 200   b 5%
   c 40%
3 a $\frac{4}{52} = \frac{1}{13}$ b $\frac{4}{52} = \frac{1}{13}$
   c $\frac{8}{52} = \frac{2}{13}$ d $\frac{48}{52} = \frac{12}{13}$
   e $\frac{44}{52} = \frac{11}{13}$

Page 25

1 a 36
   b i $\frac{6}{36} = \frac{1}{6}$ ii $\frac{4}{36} = \frac{1}{9}$
   c i $\frac{4}{36} = \frac{1}{9}$ ii $\frac{6}{36} = \frac{1}{6}$ iii 7
2 a $\frac{5}{12}$ b $\frac{1}{12}$
   c $\frac{5}{12}$

Page 26

1 a $\frac{1}{16}$ b $\frac{7}{16}$
2 a $\frac{1}{15}$ b $\frac{2}{5}$
   c $\frac{8}{15}$

Page 28

1 a 35   b £11.34
2 a 18   b 13
   c 12
3 a The last digit should be 5 not 8
   b 120 is less than half of 250 so the answer
      should be under 0.5

Page 29

1 a i 4   ii 0.8   iii 60
   b i 57 ii 0.97 iii 89

2 a 700   b 5
   c 10   d 24
3 a 300   b 760
   c 0.075   d 0.034
   e 600 000   f 200

Page 30

1 a i 30   ii 84   iii 130   iv 36
       v 40   vi 300
   b i $2^2 \times 5$ ii $3^2 \times 5$ iii 2^6 iv $2^3 \times 3 \times 5$
2 a i 30   ii 21   iii 39
   b The LCM is the product of the two numbers
   c i 18 ii 40 iii 75
   d i 6 ii 2 iii 5 iv 16 v 12 vi 1

Page 31

1 a $\frac{2}{5}$ b $\frac{3}{4}$
   c $\frac{1}{3}$
2 a i $\frac{19}{28}$ ii $1\frac{5}{18}$ iii $6\frac{1}{15}$
   b i $\frac{13}{30}$ ii $\frac{2}{9}$ iii $\frac{7}{12}$
   c i $\frac{1}{6}$ ii $\frac{5}{14}$ iii $3\frac{17}{20}$
   d i $\frac{7}{10}$ ii $1\frac{3}{4}$ iii $1\frac{3}{5}$

Page 32

1 a i 0.8   ii 0.07   iii 0.22
   b i 1.05 ii 1.12   iii 1.032
   c i 0.92 ii 0.85   iii 0.96
2 a £168   b 66.24 kg
3 a i 740   ii 5%
   b 20%

Page 33

1 a £614.63   b 1371
2 a £350   b 14 000

Page 34

1 a £100 and £400   b 50 g and 250 g
   c £150 and £250   d 80 kg and 160 kg
2 a 56   b 30
3 a 37.5 mph   b 52.5 km

Page 35

1 a 31.25 kg   **b** 160

2 a travel-size (75 g) – 1.44 g/p compared to 1.38 g/p for large tube (250 g)

  b 95 out of 120 – 79.2% compared to 62 out of 80 which is 77.5%

3 a 6.25 g/cm^3 **b** 27 kg

  c 0.3 m^3

Page 36

1 a i 9 **ii** 4 **iii** 5

  b i ±2   **ii** +1   **iii** −2

  c i 24   **ii** 1.1   **iii** 6.1

2 a i 27   **ii** 64   **iii** 1000

  b i 4^5 **ii** 6^6 **iii** 10^4 **iv** 2^7

  c i 1024   **ii** 46 656   **iii** 10 000   **iv** 128

  d 64, 128, 256, 512, 1024

3 a i $\frac{1}{64}$ **ii** $\frac{1}{7}$ **iii** $\frac{1}{9}$

  b i $\frac{2}{3}$ or 8^{-1} **ii** 3^{-1} **iii** x^{-n}

Page 37

1 a i 2^7 **ii** 2^9 **iii** x^9 **iv** 3^3 **v** 3^4 **vi** x^4

  b i $4a^2b^2c^3$ **ii** $4x^2y^3z$

2 a x^{15} **b** $27a^6$

  c $4x^2y^4$

3 a 5   **b** 4

  c $\frac{1}{5}$

4 a $\frac{1}{16}$ **b** $\frac{1}{9}$

  c $\frac{1}{100\,000}$

Page 38

1 a i 5.6×10^5 **ii** 7×10^{-6} **iii** 3×10^6

  b i 6 400 000   **ii** 0.000 83   **iii** 900 000 000

  c i 1.56×10^7 **ii** 4×10^3

    iii 9×10^6 **iv** 8×10^{-5}

2 a $0.\dot{3}\dot{6}$ **b** $0.\dot{6}1\dot{5}$

  c $0.3\dot{6}$ **d** $0.\dot{6}$

  e $0.1\dot{6}$ **f** $0.\dot{7}$

3 a i 0.25   **ii** 0.05   **iii** $0.\dot{1}$

  b i $1\frac{1}{7}$ **ii** $1\frac{4}{5}$ **iii** $4\frac{1}{3}$

Page 39

1 a 2   **b** 36

  c $5\sqrt{3}$ **d** $9\sqrt{2}$

  e $3 + 2\sqrt{3}$ **f** −1

  g $14 - 6\sqrt{5}$

2 $7 + 3\sqrt{3}$

3 a $\frac{\sqrt{3}}{2}$ **b** $\frac{3}{2}$

  c $\frac{3\sqrt{5} - 5}{5}$

4 $ad \div bc$ must be a square number.

$$\sqrt{\frac{a}{b}} \div \sqrt{\frac{c}{d}} = N \Rightarrow \sqrt{\frac{a \div b}{c \div d}} = N \Rightarrow \sqrt{\frac{ad}{bc}} = N \Rightarrow \frac{ad}{bc} = N^2$$

Page 40

1 a $y = kx^2$ **b** $\frac{1}{20}$ or 0.05

  c 1.25   **d** 20

2 a $y = \frac{k}{\sqrt{x}}$ **b** 12

  c 4   **d** 1

Page 41

1 a 6.5 m, 7.5 m

  b 33.5 kg, 34.5 kg

  c 315 cm, 325 cm

2 a 86.25 cm^2 ≤ area < 106.25 cm^2

  b 29.1 mph ≤ speed < 35.5 mph

Page 43

1 a 37.7 cm   **b** 8π cm

2 a 706.9 cm^2 **b** 9π cm^2

3 a 37.5 cm^2 **b** 34 cm^2

Page 44

1 a i 24 cm^2 **ii** 240 cm^3

  b 42 m^3

2 a arc length = 8.7 cm

    area = 43.6 cm^2

  b i $\frac{4}{3}\pi + 12$ cm   **ii** 4π cm^2

Page 45

1 160π cm^3

2 112π cm^2

3 605 cm^3

4 288.7 cm^3

Page 46

1 i 301.6 cm³   ii 188.5 cm²

2 i 36π cm³   ii 36π cm²

3 i 261.8 cm³   ii 235.6 cm²

4 Ratio of length of small cone to large cone = 1 : 2
   Ratio of volume of small cone to large cone = 1 : 8
   Ratio of volume of small cone to frustum = 1 : 7

Page 47

1 $3^2 + 4^2 = 5^2$

2 a 11.2 cm   b 3.9 m

3 18.4 cm

Page 48

1 36.7 cm²

2 7.07 cm

Page 49

1 a i 0.515   ii 0.574   iii 0.176
   b i 36.9°   ii 45.6°   iii 63.4°

2 30°; 8.34 cm; 7.52 cm

Page 50

1 a 5.32 cm   b 32.2°
   c 11.3 cm

2 26.0 m

Page 51

1 a 5.14 cm   b 6.86 cm
   c 41.8°

2 35.3°

Page 52

1 a 18.9°   b 54.2°

2 a 25.8°   b 10.9 cm

Page 53

1 a 23.6° and 156.4°
   b 84.3° and 275.7°
   c 224.4° and 315.6°
   d 113.6° and 246.4°

Page 54

1 a 69°   b 12.6 cm

2 55.7° and 124.3°

Page 55

1 7.5 cm

2 112.4°

Page 56

1 a 10.6 cm   b 16.0 cm

2 a 95.1°   b 126.5°

3 33.8 cm²

4 a Use the sine rule to work out angle A
     Use angles in a triangle to work out angle B
     Use the rule $\frac{1}{2}ac \sin B$ to work out the area
   b 33.0 cm²

Page 57

1 a yes – a square has all the properties of a
     rectangle
   b yes – a rhombus has all the properties of a
     parallelogram
   c no – a kite does not have all sides equal

2 a i 135°   ii 45°
   b i 140°   ii 40°

3 They always add up to 180°.

Page 58

1 a 80° – opposite angle in a cyclic quadrilateral
   b 160° – angle at centre twice angle at
     circumference

2 a 58° – sum of angles in a triangle
   b 32° – angle in same segment

Page 59

1 15° – angles in triangle are $5x + 5x + 2x = 180$

2 62° – angle in alternate segment is 56° and
   other two angles are equal (isosceles triangle)

Page 60

1 **a** A and C – SAS

   b B and D – D is 1.25 times bigger than B

2 **a** $\binom{5}{0}$ **b** $\binom{3}{4}$

   c $\binom{5}{-4}$ **d** $\binom{0}{-4}$

3 Yes, as the angles are the same and the sides are in the same ratio, 1 : 1

Page 61

1 **a** y-axis   **b** $x = 3$

2 **a** 90°, anticlockwise about (0, 0)

   b half-turn about (–1, 0)

Page 62

1 **a i** 2   **b ii**

   b i $\frac{2}{3}$

   a ii

2 **a** reflection in y-axis

   b reflection is x-axis

   c 180° rotation about (0, 0)

Page 63

1 **a–c** self-checking

2 accurate drawing

Page 64

1–3 self-checking

Page 65

1 self-checking

2

•Edinburgh

•London

Page 66

1 20 m

Page 67

1 **a** 3   **b** 9

   c 27

2 18.5 kg

Page 68

1 **a i** a + b   **ii** 3a   **iii** 2a + 2b

   iv –2a + b   **v** –a – 2b

   b $\overrightarrow{AT}$ is three times $\overrightarrow{BH}$ and parallel to it.

   c They are the same magnitude and in opposite directions

   d They lie on a straight line as $\overrightarrow{OD}$ and $\overrightarrow{DS}$ are parallel and have a common point

Page 69

1 **a i** –a + b or b – a

   ii $\frac{1}{3}$a + $\frac{1}{3}$b

   iii –$\frac{1}{2}$a + b

   b $\overrightarrow{AP}$ = –a + 2b which is 2$\overrightarrow{AM}$, hence $\overrightarrow{AP}$ and $\overrightarrow{AM}$ are multiples of each other and share a common point, so AMP is a straight line

Page 72

1 **a** –7   **b** 25

   c –28

2 **a** $3x + 15$ **b** $n^2 - 7n$

   c $6p^3 - 9p^2q$

3 **a** $6x + 6$ **b** $10x + 38y$

4 **a** $3x(2x - 3)$ **b** $2ab(a - 4 + 3b)$

Page 73

1 **a** $x = 3$ **b** $m = -1$

   c $n = 12$ **d** $x = \frac{1}{2}$

   e $x = 13$ **f** $x = 2$

   g $y = 13$ **h** $x = 28$

   i $x = \frac{1}{2}$

2 **a** $x = 9$ **b** $m = 6$

   c $x = -2$ **d** $x = 4$

   e $x = 2\frac{1}{2}$ **f** $y = -1\frac{1}{2}$

3 **a** $x = 7$ **b** $y = 2$

   c $x = 3$ **d** $x = -1$

   e $x = \frac{1}{5}$ **f** $y = -2\frac{1}{2}$

Page 74

1 **a** $x = 9$ **b** $x = -3$

   c $x = 5$ **d** $x = 6\frac{1}{2}$

2 **a** 4   **b** 6

3 3

4 **a** $\frac{x}{2} + 7 = x + 6$ **b** 2

Page 75

1 4.8

2 2.6

Page 76

1 **a** $x = 3, y = -2$

b $x = 2\frac{1}{2}, y = 1\frac{1}{2}$

c $x = 5, y = 1$

2 $x = 4, y = -1$

3 When you subtract the equations you get $0 = 5$

Page 77

1 **a** £7.50　　　　**b** £5

2 **a** $x = \frac{T}{4}$　　　　**b** $x = \frac{y-3}{2}$

c $x = 5y$

3 **a** $x = P - 2t$　　　**b** $x = \frac{A-y}{m}$

c $x = \sqrt{\frac{S}{2\pi}}$

Page 78

1 **a** $x^2 + 2x - 3$　　**b** $m^2 - 4m - 12$

c $n^2 - n - 2$　　**d** $x^2 + 6x + 5$

e $x^2 - 6x + 9$　　**f** $x^2 + 10x + 21$

2 **a** $x^2 + 10x + 25$　**b** $x^2 - 4x + 4$

3 **a** $x^2 - 1$　　　　**b** $m^2 - 4$

Page 79

1 **a** $(x + 3)(x - 1)$　　**b** $(x - 3)(x - 2)$

2 **a** $(x + 4)(x - 4)$　　**b** $(x + 6)(x - 6)$

3 **a** $(2x + 1)(x - 5)$　　**b** $(3x - 2)(2x + 3)$

4 **a** $x = -1$ or $x = 3$　**b** $x = -5$ or $x = 2$

Page 80

1 **a** $x = -2$ or $x = \frac{3}{2}$　**b** $x = -\frac{3}{4}$ or $x = -\frac{2}{3}$

2 **a** $x = \pm\sqrt{\frac{7}{2}}$　　　**b** $x = 0$ or $x = 3$

3 $x = -0.28$ or $x = 1.78$

4 $2x^2 - 5x - 4 = 0$

Page 81

1 **a** $(x + 3)^2 - 10$　　**b** $(x - 4)^2 - 19$

2 **a** $-2 \pm \sqrt{5}$　　　**b** $3 \pm \sqrt{7}$

3 **a** because $b^2 - 4ac = -4 < 0$, it has no solutions

b $-3 \pm \sqrt{19}$

Page 82

1 **a** $\frac{3 \pm \sqrt{13}}{2}$　　　　　　　**b** $\frac{-3 \pm \sqrt{11}}{2}$

2 **a** $x - \frac{6}{x} = 1 \Rightarrow x^2 - 6 - x = 0$

b $x = -2$ or $x = 3$

3 2 m

4 $x = -\frac{1}{2}$ and $-\frac{11}{2}$

Page 83

1 **a** 20 minutes　　　**b** car was stopped

c B and C　　　　**d** 45 km/h

2 **a**

b

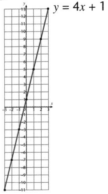

Page 84

1

$y = 4x + 1$

2 4

Page 85

1 **a** A and C　　　　**b** A and B

2 **a** 　$y = 4x + 3$　**b** 　$y = \frac{1}{2}x + 1$

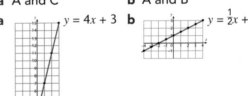

3 a   **b**

4 A as it does not pass through the point (0, 4) *or* C as A and B have the same gradient

Page 86

1 $y = 3x + 4$

2 a £6.50　　　　　**b** $1\frac{1}{2}$

　c 2　　　　　　　**d** $C = 1\frac{1}{2}d + 2$

　e £6.50

Page 87

1 $x = 2, y = 3$

2 $y = -3x + 5$

3 $y = 2x - 3$

Page 88

1 a (–3, 5), (–2, 1), (–1, –1), (0, –1), (1, 1), (2, 5), (3, 11)

　b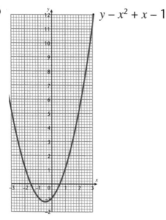
$y - x^2 + x - 1$

　c –2.3, 1.3

　d –1.6, 0.6

Page 89

1 a (0, –2)

　b (–1.5, –4.25)

　c 0.6, –3.6

　d line is $y = -1$, solutions are 0.3, –3.3

　e line is $y = x$, solutions are 0.7, –2.7

Page 90

1 Table: 0, 0.38, 0, 13.13, 24

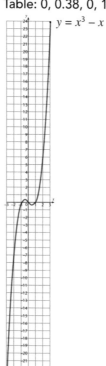
$y = x^3 - x$

2 Table: 5.66, 8, 11.3

$y = 2^x$

Page 91

1 a $\frac{5x - 1}{6}$　　　　　**b** $\frac{-7x - 5}{15}$

　c $\frac{3xy}{10}$　　　　　　**d** $\frac{4}{3}$

2 a $\frac{2x + 5}{35}$　　　　　**b** $\frac{4xy^2}{15}$

　c 1

3 a $\frac{3x - 19}{x^2 - x - 6}$　　　**b** $\frac{x^2 + x + 6}{x^2 + 2x - 8}$

Page 92

1 a $x = 5$　　　　　**b** $x = -\frac{3}{2}$ or $x = 3$

2 a $\frac{x + 1}{2x + 1}$　　　　　**b** $\frac{3x - 1}{2x + 3}$

3 $x^2 - 2x - 3$ and $2x + 1$

Page 93

1 a (4, 3), (–3, –4)　　**b** (3, 2), (–2, 3)

　c $3\frac{1}{2}$ and –2　　　**d** $-\frac{1}{2}$ and $-\frac{2}{3}$

Page 94

1 a i 4, 9, 14 **ii** 499

 b i 3, 6, 10 **ii** 20 100

2 a $5n + 1$ **b** $8n - 5$

 c $3n + 6$

3 a 200 **b** 199

 c 10 000 **d** 28, 36, 45, 55

 e 128, 256, 512, 1024

 f 23, 29, 31, 37, 41

4 The first sequence increases by 5 and the second by 10, which is a multiple of 5 so as they have different starting terms, they will never have a common term

Page 95

1 a 41 **b** $4n + 1$

2 a $x = \frac{2y}{a - 2}$ **b** $x = \frac{4 + 2y}{y - 1}$

3 $(n + 2)^2 - 9 = n^2 + 4n + 4 - 9 = n^2 + 4n - 5$

 $(n + 2 - 3)(n + 2 + 3) = (n - 1)(n + 5) = n^2 + 4n - 5$

Page 96

1 a $x < 3$ **b** $x > 1$

 c $x \geq 18$ **d** $x > -10$

2 a $x > 1$ **b** $x \leq 3$

3 a $x < 2$ **b** $x \geq -1$

 c $-1, 0, 1$

4 $-1, 0, 1, 2, 3, 4, 5$

Page 97

1 a **b** **c**

2

Page 98

1

2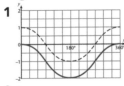

3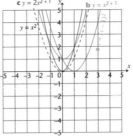

Page 99

1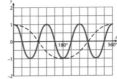

2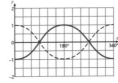

3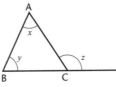

Page 100

1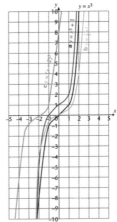

Using triangle ABC, angle ACB $= 180 - (x + y)$

Angle ACB $= 180 - z$ (angles on a straight line)

Hence $180 - (x + y) = 180 - z$

Hence $x + y = z$

2 $(n + 6)^2 - (n + 2)^2 = n^2 + 12n + 36 - (n^2 + 4n + 4)$

 $= n^2 + 12n + 36 - n^2 - 4n - 4$

 $= n^2 - n^2 + 12n - 4n + 36 - 4$

 $= 8n + 32 = 8(n + 4)$

3 $AC = CB$ (side of an isosceles triangle)

 CM is a common side

 $AM = MB$ (M midpoint of base)

 Congruent due to SSS

NEW GCSE
MATHS
Higher

Exam Practice Workbook

For GCSE Maths from 2010

Edexcel + AQA + OCR

Keith Gordon

D

1 The table shows the number of cars per house on a housing estate of 100 houses.

Number of cars	Number of houses
0	8
1	23
2	52
3	15
4	2

  a Find the modal number of cars.

                                                             [1 mark]

  b Find the median number of cars.

                                                             [1 mark]

  c Calculate the mean number of cars.

                                                         [3 marks]

C

2 The table shows the results of a survey in which 50 boys were asked how many brothers and sisters they had.

		Number of brothers				
		0	1	2	3	4
	0	6	8	3	0	0
	1	3	5	7	1	0
Number of	2	1	2	6	2	0
sisters	3	1	0	1	2	1
	4	0	0	1	0	0

  a What is the modal number of sisters?   **[1 mark]**

  b What is the median number of sisters?   **[1 mark]**

  c Calculate the mean number of brothers.   **[3 marks]**

  (AU d) Mary says that the boys in this survey have more brothers than sisters. Explain why Mary is correct.

                                                                   [1 mark]

C

(PS 3) Find five numbers with a mean of 8, a mode of 9 and a range of 4.

                                                                    [3 marks]

FM 1 The table shows the scores of 200 boys in a mathematics examination.

The frequency polygon shows the scores of 200 girls in the same examination.

Mark, x	Frequency, f
$40 < x \leqslant 50$	27
$50 < x \leqslant 60$	39
$60 < x \leqslant 70$	78
$70 < x \leqslant 80$	31
$80 < x \leqslant 90$	13
$90 < x \leqslant 100$	12

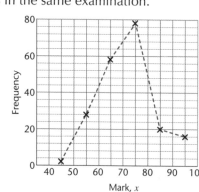

a Calculate the mean score for the boys.

_____ **[3 marks]**

b Draw the frequency polygon for the boys' scores on the same axes as the girls' scores. **[1 mark]**

c Who did better in the test, the boys or the girls? Give reasons for your answer.

_____

_____ **[1 mark]**

2 The table shows the lengths of 100 cucumbers.

Length, x (cm)	Frequency	Frequency density
$10 < x \leqslant 20$	15	
$20 < x \leqslant 25$	24	
$25 < x \leqslant 30$	36	
$30 < x \leqslant 40$	18	
$40 < x \leqslant 60$	7	

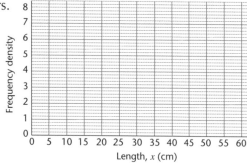

a Complete the table by calculating the frequency density for each class interval. **[2 marks]**

b Draw a histogram to show this information. **[2 marks]**

PS 3 The histogram shows the ages of members of a bowls club.

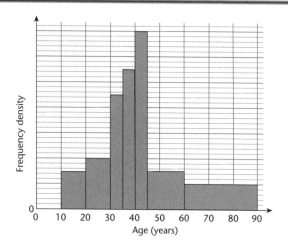

There are 115 members under 35.

How many members are over 55?

_____ **[4 marks]**

FM 1 Danny wanted to investigate the hypothesis: *'Girls spend more time on their mathematics homework than boys do.'*

a Design a two-way table that Danny can use to record his data.

[2 marks]

b Danny collected data from 30 boys and 10 girls.

He found that, on average, the boys spent 10 hours on their mathematics homework and the girls spent 11 hours on theirs. Does this prove the hypothesis?

Give reasons for your answer.

_____

_____ **[1 mark]**

FM 2 A farmer thinks that the amount of rainfall is increasing each year.

He records the amount of rain that falls each month for a year.

The table shows the results (in mm).

	Jan	Feb	Mar	Apr	May	Jun	Jul	Aug	Sep	Oct	Nov	Dec
2009	39	45	32	38	56	22	18	21	34	39	45	43

On the internet he finds out that the mean rainfall for 2008 is 35 mm, with a range of 40.

Investigate the hypothesis: *'The amount of rainfall is increasing each year.'*

[5 marks]

AU 3 Describe the method of data collection you would use to investigate the following.

A   The likelihood of an earthquake in Hawaii

_____

B   The fairness of a home-made spinner

_____

C   The probability of winning a prize in a raffle

_____

D   The percentage of people in a small town who would like to see a new supermarket opened.

_____

[3 marks]

1   Daman does a survey about school lunches.

   a   This is one of the questions in his survey.

   Burgers and chips are popular but not healthy. Don't you agree? ☐ Yes   ☐ No

   Give two criticisms of this question

   Criticism 1 _____

   Criticism 2 _____

[2 marks]

   b   This is another of his questions.

   How much do you spend, on average, on school lunch?

   Design a response section for this question.

[2 marks]

FM 2   A national grocery chain wants to build a supermarket in a small market town.

   One morning they asked 50 people who were shopping on the high street whether they ever shopped in supermarkets. 32 of them said that they did.

   At the planning meeting, the grocery chain claimed that 7 out of 10 of the people in the town wanted the supermarket.

   a   Give two criticisms of the sampling method.

   Criticism 1 _____   **[1 mark]**

   Criticism 2 _____   **[1 mark]**

   b   Why is the conclusion invalid?

   _____   **[1 mark]**

AU 3   A transport company manager wants to survey her staff about working conditions.

This is the breakdown of the employees, by gender and job.

	Managers	Drivers	Maintenance
Male	8	182	56
Female	5	89	26

She decides to survey 50 staff.

	Managers	Drivers	Maintenance
Male	1	25	8
Female	1	12	4

The manager works out that she will need to survey this many people in each category.

Comment on her decisions.

_____

_____   **[3 marks]**

D

FM1 This line graph shows the temperature in a greenhouse throughout one morning.

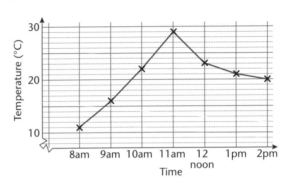

a   Between which two hours was the biggest increase in temperature? _____ **[1 mark]**

b   The gardener opened the ventilator to lower the temperature. At what time did he do this? _____ **[1 mark]**

c   What was the approximate temperature at 10:30am? _____ **[1 mark]**

d   Can you use the graph to predict the temperature at 6pm? Explain your answer.

_____

_____ **[1 mark]**

C

FM2 A teacher recorded how many times the students in her form were late during a term.

The stem-and-leaf diagram shows the data.

12 students were **never late**.

```
0 | 2  3  4  4  5  6  7
1 | 3  5  8  9  9
2 | 0  1  4  5
3 | 2
5 | 1
Key:  1 | 7 represents 17
```

a   How many students were in her form altogether? _____ **[1 mark]**

b   Work out the mean number of lates for the **whole form**. _____ **[3 marks]**

C

PS3 This graph shows the price index of petrol from 2000 to 2007. In 2000 the price of a litre of petrol was 60p.

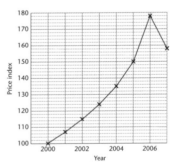

a   Use the graph to find the price of a litre of petrol in 2007. _____ **[1 mark]**

b   One year there was an oil crisis as a result of a war in Iraq. What year do you think it was? Give a reason for your answer.

_____ **[1 mark]**

c   Over the same seven-year period, the index for the cost of living rose by 38%.

Did the cost of petrol increase more quickly or more slowly than the cost of living? Explain your answer.

_____ **[1 mark]**

FM 1 A delivery driver records the distances and times for deliveries.

The table shows the results.

Delivery	Distance (km)	Time (minutes)
A	12	30
B	16	42
C	20	55
D	8	15
E	18	40
F	25	60
G	9	45
H	15	32
I	20	20
J	14	35

a Plot the values on the scatter diagram.

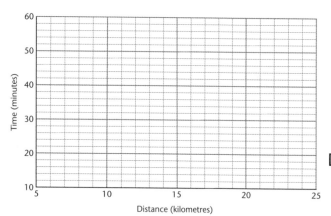

[2 marks]

b i During one of the deliveries the driver was stuck in a traffic jam.

Which delivery was this? _____ **[1 mark]**

ii One of the deliveries was done very early in the morning when there was no traffic. Which delivery was this? _____ **[1 mark]**

c Ignoring the two values in **b**, draw a line of best fit through the rest of the data. **[1 mark]**

d Under normal conditions, how long would you expect a delivery of 22 kilometres to take? _____ **[1 mark]**

AU 2 Match each of these scatter diagrams to one of the statements below.

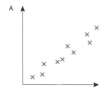 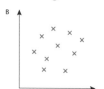  

W: How much is spent on school lunch against distance lived from the school _____

X: Percentage value of the initial cost of a car against the age of the car _____

Y: Marks in a mathematics exam against marks in a science exam _____

Z: Cost of a taxi journey against distance _____

[2 marks]

C

Cumulative frequency and box plots

B **FM 1** The cumulative frequency diagram shows the amount of time 80 shoppers waited in a supermarket queue before being served.

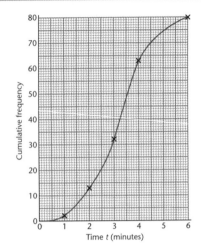

a What is the median queuing time? _____ **[1 mark]**

b What is the interquartile range? _____ **[2 marks]**

c Use the graph to estimate the number of customers who spent less than $3\frac{1}{2}$ minutes in the queue. _____ **[1 mark]**

B **FM 2** The box plots show the marks students in two forms, 10K and 10J, achieved in the same geography test.

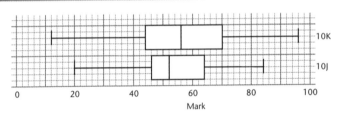

a What was the lowest mark in form 10K? _____ **[1 mark]**

b What was the median mark for form 10J? _____ **[1 mark]**

c Which form did better in the test? Give a reason for your answer.

_____ **[1 mark]**

d Which form was more consistent in the test? Give a reason for your answer.

_____ **[1 mark]**

B **AU 3** Below are four cumulative frequency diagrams and four box plots.

Match each cumulative frequency diagram with a box plot.

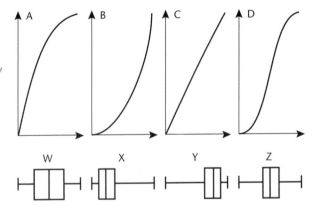

A: _____ B: _____ C: _____ D: _____ **[2 marks]**

1 John made a dice and weighted one side with a piece of sticky gum.

He threw it 120 times. The table shows results.

Score	1	2	3	4	5	6
Frequency	18	7	22	21	35	17
Relative frequency						

a Complete the table by calculating the relative frequency of each score.

Give your answers to 2 decimal places. **[2 marks]**

b On which side did John stick the gum?

Give a reason for your answer. _____ **[1 mark]**

2 A bag contains 30 balls that are either red or white. The ratio of red balls to white balls is 2 : 3.

a Explain why the events 'picking a red ball at random' and 'picking a white ball at random' are mutually exclusive.

_____

_____ **[1 mark]**

b Explain why the events 'picking a red ball at random' and 'picking a white ball at random' are exhaustive.

_____

_____ **[1 mark]**

c Zoe says that the probability of picking a red ball at random from the bag is $\frac{2}{3}$. Explain why Zoe is wrong.

_____

_____ **[1 mark]**

d How many red balls are there in the bag? _____ **[1 mark]**

e A ball is drawn at random from the bag, its colour is noted and then it is replaced in the bag.

This is repeated 200 times. How many of the balls would you expect to be red?

_____ **[1 mark]**

PS 3 Three bags contain white and black balls.

Bag A contains three white and three black balls.

Bag B contains three white and four black balls.

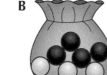

Bag C contains four white and five black balls.

Two bags are to be combined together.

Which two bags should be combined to give the greatest chance of picking a white ball?

_____ **[3 marks]**

1   The two-way table shows the gender and departments of 40 teachers in a school.

	Male	Female
Mathematics	7	5
Science	11	7
RE	1	3
PE	3	3

a   How many male teachers are there?   _____   **[1 mark]**

b   Which subject has equal numbers of
   male and female teachers?   _____   **[1 mark]**

c   Nuna says that science is a more popular subject than mathematics for female teachers.

   Explain why Nuna is wrong.

   _____

   _____   **[1 mark]**

d   What is the probability that a teacher chosen at random will be a female teacher of
   mathematics or science?

   _____   **[1 mark]**

2   The table shows the probability, by gender, of right- or left-handedness
   of a student picked at random from a school.

	Boys	Girls
Right-handed	0.36	0.4
Left-handed	0.1	0.14

a   What is the probability that a student picked at random
   from the school is left-handed?   _____   **[1 mark]**

b   There are 432 right-handed boys in the school.

   How many students are there in the school altogether?   _____   **[2 marks]**

c   In the town that the school serves, there are 26 000 people.

   Estimate how many of them are left-handed.   _____   **[2 marks]**

PS 3   Here is some information about an exercise class of 25 people.

   • There are five more women than men

   • A third of the women are 65 or over.

   • Four times as many men are under 65 than 65 or over.

   Use the information to fill in the two-way table.

	Men	Women	Total
65 or over			
Under 65			
Total			

[3 marks]

1 A fair 4-sided spinner is spun twice.

If the two scores are odd, the scores are added.

If one or both of the scores is even, the two scores are multiplied.

a Complete the table.

		First score			
		1	2	3	4
	1	2	2	4	
Second	2				
score	3				
	4				

[2 marks]

b What is the probability that the combined score will be an even number?

_____ [1 mark]

c What is the probability that the combined score will be a square number?

_____ [1 mark]

2 A bag contains three red balls and seven blue balls. A ball is taken out at random and replaced.

Another ball is then taken out.

a Complete the tree diagram to show the probabilities of the possible outcomes.

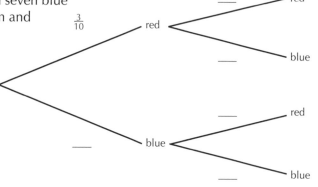

[1 mark]

b What is the probability that both balls are red? _____ [2 marks]

c What is the probability that the balls are different colours? _____ [2 marks]

(AU 3) A bag contains some white balls and some black balls.

A ball is taken out at random and its colour noted.

The ball is then replaced in the bag.

Another ball is then taken out at random and its colour noted.

a Which of these **could not** be the probability of taking two white balls.

$\frac{4}{25}$ $\frac{9}{16}$ $\frac{16}{19}$

Give a reason for your choice. _____ [1 mark]

b It is known that there are more black balls than white balls in the bag.

Which of the probabilities in part **a** must be the probability of taking two white balls?

Give a reason for your choice. _____ [1 mark]

Statistics

B

1 A bag contained red and blue counters in the ratio 1 : 4.

    a What was the probability of picking a red counter at random from the bag?

        _____ **[1 mark]**

    b There were six red counters in the bag. How many blue counters were there?

        _____ **[1 mark]**

   (PS c) Some red counters were added to the bag so that the probability of picking a red counter was $\frac{1}{2}$.

    How many red counters were added?   _____ **[1 mark]**

A

2 The probability of throwing a 6 with a biased dice is x.

    a What is the probability of **not** throwing a 6?   _____ **[1 mark]**

    b The dice is thrown three times.

        What is the probability of throwing three 6s?   _____ **[1 mark]**

A

3 A box contains two brown eggs and four white eggs.
Mandy uses two eggs to make an omelette.

    a Complete the tree diagram to show
the probabilities of the possible
outcomes.

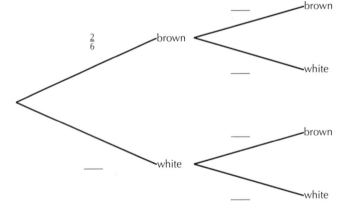

        [2 marks]

    b What is the probability she uses two eggs of the same colour? _____ **[2 marks]**

A*

(PS 4) A bag contains x blue balls and y white balls.

    The probability of taking a blue ball at random from the bag is p.

    Five blue balls are taken from the bag and discarded.

    The probability of taking a white ball from the bag is now p.

    a Show that $x^2 - 5x = y^2$.

        _____ **[2 marks]**

    b Use trial and improvement to find the smallest possible values of x and y.

        _____ **[2 marks]**

Statistics record sheet

Name _____   Marks _____

Form _____   Percentage _____

Date _____   Grade _____

Page number	Question number	Topic	Mark	Comments
112	1	Statistics 1	/5	
	2	Statistics 1	/6	
	3	Statistics 1	/3	
113	1	Statistics 1	/5	
	2	Statistics 1	/4	
	3	Statistics 1	/4	
114	1	Statistics 1	/3	
	2	Statistics 1	/5	
	3	Statistics 1	/3	
115	1	Statistics 1	/4	
	2	Statistics 1	/3	
	3	Statistics 1	/3	
116	1	Statistics 2	/4	
	2	Statistics 2	/4	
	3	Statistics 2	/3	
117	1	Scatter diagrams	/6	
	2	Scatter diagrams	/2	
118	1	Cumulative frequency and box plots	/4	
	2	Cumulative frequency and box plots	/4	
	3	Cumulative frequency and box plots	/2	
119	1	Probability	/3	
	2	Probability	/5	
	3	Probability	/3	
120	1	Probability	/4	
	2	Probability	/5	
	3	Probability	/3	
121	1	Probability	/4	
	2	Probability	/5	
	3	Probability	/2	
122	1	Probability	/3	
	2	Probability	/2	
	3	Probability	/4	
	4	Probability	/4	
Total			**/124**	

Successes

1 _____

2 _____

3 _____

Areas for improvement

1 _____

2 _____

3 _____

D **FM1** **a** The school canteen has 34 tables that each seat 14 students and 12 tables that each seat 8 students. What is the maximum number of students that can sit in the canteen at the same time?

_____ **[2 marks]**

PS b The head wants to expand the canteen so that it can cater for 700 students.

He wants to purchase an equal number of 12-seater and 8-seater tables.

How many extra tables will he need?

_____ **[2 marks]**

D **FM2** **a** Work these out.

   i $10.8 \div 0.6$

_____ **[1 mark]**

   ii $8.64 \div 0.36$

_____ **[1 mark]**

b A dividend is a repayment made every few months on the amount spent.

The Co-op pays a dividend of 2.6 pence for every £1 spent.

   i In three months, Derek spends £240. How much dividend will he get?

_____ **[1 mark]**

   ii Doreen received a dividend of £7.80 How much did she spend to get this dividend?

_____ **[1 mark]**

C **PS3** This year I am twice as old as my daughter.

Fourteen years ago I was four times as old as she was then.

How old am I now?

_____ **[3 marks]**

1 a Round each number to the number of significant figures (sf) indicated.

   i   67 800 (1 sf)

                                                               _____ [1 mark]

   ii 0.067 42 to (2 sf)

                                                               _____ [1 mark]

  b Find approximate answers to the following.

   i $\dfrac{312 \times 7.92}{0.42}$

                                                               _____ [2 marks]

   ii $\dfrac{487}{0.52 \times 3.98}$

                                                               _____ [2 marks]

2 a Write down the answers to:

   i 3.7×100 _____ [1 mark]

   ii 0.25×10^3 _____ [1 mark]

  b Write down the answers to:

   i $7.6 \div 10$ _____ [1 mark]

   ii $0.65 \div 10^2$ _____ [1 mark]

  c Write down the answers to:

   i $30\,000 \times 400$ _____ [1 mark]

   ii 600^2 _____ [1 mark]

  d Write down the answers to:

   i $90\,000 \div 30$ _____ [1 mark]

   ii $30\,000 \div 60$ _____ [1 mark]

(AU 3) Here are three decimals.

  A 3.65   B 3.725   C 0.3627

  a Give a mathematical property that A and B have in common.

                                                              _____ [1 mark]

  b Give a mathematical property that B and C have in common.

                                                              _____ [1 mark]

  c Which is larger, $3.65 \div 10$ or 0.3627?

                                                              _____ [1 mark]

C

1 **a** You are told that p and q are prime numbers.

$p^2q^2 = 36$

What are the values of p and q?

$p =$ _____

$q =$ _____ **[2 marks]**

b Write 360 as the product of its prime factors.

_____ **[2 marks]**

c You are told that a and b are prime numbers.

$ab^2 = 98$

What are the values of a and b?

$a =$ _____

$b =$ _____ **[2 marks]**

d Write 196 as the product of its prime factors.

_____ **[1 mark]**

C

2 **a** Write 24 as the product of its prime factors.

_____ **[1 mark]**

b Write 60 as the product of its prime factors.

_____ **[1 mark]**

c What is the lowest common multiple of 24 and 60?

_____ **[1 mark]**

d What is the highest common factor of 24 and 60?

_____ **[1 mark]**

e In prime factor form, the number $P = 2^4 \times 3^2 \times 5$ and the number $Q = 2^2 \times 3 \times 5^2$.

i What is the lowest common multiple of P and Q?

Give your answer in index form.

_____ **[1 mark]**

ii What is the highest common factor of P and Q?

Give your answer in index form.

_____ **[1 mark]**

(AU 3) a, b and c are prime numbers.

X and Y are two numbers such that

$X = a^2bc$ and $Y = ab^2c^3$

a Circle the expression below that represents the LCM of X and Y.

a^2bc^3 $a^2b^2c^3$ a^2bc abc

b Circle the expression below that represents the HCF of X and Y.

a^2bc^3 $a^2b^2c^2$ a^2bc abc

[2 marks]

Fractions

1 **a** Work out $\frac{3}{4} + \frac{2}{5}$.

Give your answer as a mixed number. _____ **[2 marks]**

b Work out $3\frac{2}{3} - 1\frac{4}{5}$.

Give your answer as a mixed number. _____ **[3 marks]**

c On an aeroplane, two-fifths of the passengers are British, one-quarter are German, one-sixth are American and the rest are French.

What fraction of the passengers is French?

_____ **[3 marks]**

2 **a** Use your calculator to work out $2\frac{1}{2} \times 1\frac{2}{5}$.

Give your answer as a mixed number.

_____ **[1 mark]**

b Use your calculator to work out $3\frac{3}{10} \div 2\frac{2}{5}$.

Give your answer as a mixed number.

_____ **[1 mark]**

c Work out the area of this triangle.

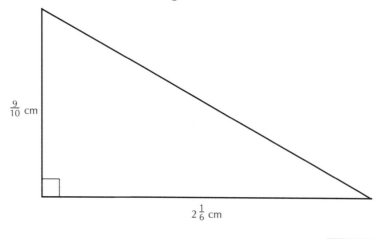

$\frac{9}{10}$ cm

$2\frac{1}{6}$ cm

_____ cm² **[2 marks]**

(**AU 3**) Here are three fractions.

$\frac{3}{2}$ $\frac{2}{3}$ $\frac{2}{8}$

Give a reason why each of them could be the odd one out.

_____

_____

_____

[3 marks]

D

FM 1 **a** A car cost £6000 new.

It depreciates in value by 12% in the first year and by 10% in the second year.

Which of these calculations shows the value of the car after 2 years?

Circle the correct answer.

6000×0.78 $6000 \times 0.88 \times 0.9$ $6000 \times 88 \times 90$ $6000 - 2200$ **[1 mark]**

b VAT is charged at 17.5 %.

A quick way to work out the VAT on an item is to work out 10%, then divide this by 2 to get 5%, then divide this by 2 to get 2.5%. Finally, add these three values together.

Use this method to work out the VAT on an item costing £68.

_____ **[2 marks]**

D

FM 2 **a** A computer costs £700 excluding VAT.

VAT is charged at 17.5%.

What is the cost of the computer when VAT is added?

_____ **[2 marks]**

b The price of a printer is reduced by 12% in a sale.

The original price of the printer was £250.

What is the price of the printer in the sale?

_____ **[2 marks]**

C

FM 3 In its first week, a new bus route carried a total of 2250 people.

In its second week it carried 2655 people.

What is the percentage increase in the number of passengers from the first week to the second?

_____ **[3 marks]**

C

AU 4 Explain why a 10% increase followed by a 10% decrease does not get back to the original amount.

_____

_____ **[2 marks]**

FM 1 **a**   A car cost £8000 new.

It depreciates in value by 12% each year.

Which of these calculations shows the value of the car after 2 years?

Circle the correct answer.

8000×0.76 $8000 \times 0.88 \times 0.88$ $8000 \times 88 \times 88$ $8000 - 2400$ **[1 mark]**

b   Mary invested £2000 in an account which paid 3.5% interest each year.

She left her money in the bank for six years.

How much money did she have at the end of this period?

_____ **[3 marks]**

c   Over the same period, Bert invests £2000 in shares which lost 2% in value each year. How much were his shares worth after six years?

_____ **[3 marks]**

FM 2 **a**   Electro Co had a dishwasher priced at £240. Its price was reduced by 5% in a sale.

Corries had the same dishwasher priced at £260. Its price was reduced by 12% in a sale.

At which company was the dishwasher cheaper?

Show all your working clearly.

_____ **[3 marks]**

b   In the same sale, Electro Co reduced the price of a cooker by 5% to £361.

What was the original price of the cooker?

_____ **[3 marks]**

AU 3   Shop A increased its prices by 5%, then reduced them by 3%.

Shop B reduced its prices by 3%, then increased them by 5%.

Which shop's percentage changes were the greatest after the changes?

Circle the correct answer and justify your choice.

Shop A   Shop B   Both the same   Cannot tell

_____ **[3 marks]**

Number

C

1 a Write the ratio 12 : 9 in its simplest form.

_____ **[1 mark]**

b Write the ratio 5 : 2 in the form 1 : n.

_____ **[1 mark]**

c A fruit drink is made from orange juice and cranberry juice in the ratio 5 : 3.

If 1 litre of the drink is made, how much of the drink is cranberry juice?

_____ **[2 marks]**

C

2 In a tutor group the ratio of girls to boys is 3 : 4. There are 15 girls in the tutor group.

How many students are there in the tutor group altogether?

_____ **[2 marks]**

C

FM 3 A ferry covers the 72 kilometres between Holyhead and Dublin in $2\frac{1}{4}$ hours.

a What is the average speed of the ferry?

State the units of your answer.

_____ **[3 marks]**

PS b For the first 15 minutes and the last 15 minutes of the journey, the ferry is manoeuvring in and out of the ports. During this time the average speed is 18 km per hour.

What is the average speed of the ferry during the rest of the journey?

_____ **[3 marks]**

C

AU 4 The ratio of pine trees to oak trees in a wood is 3 : 5.

Are the following statements True (T), False (F) or could be true (C)?

Put ticks in the appropriate boxes.

The first is done for you.

Statement	T	F	C
There are 25 pine trees in the wood.		✓	
There are 800 trees altogether in the wood.			
The fraction of pine trees in the wood is $\frac{3}{5}$.			
The percentage of oak trees in the wood is 60%.			
If half of the pine trees were cut down, the ratio of pine trees to oak trees would be 3 : 10.			

[3 marks]

FM 1 A car uses 50 litres of petrol in driving 275 miles.

    a How much petrol will the car use in driving 165 miles?

_____ **[2 marks]**

    b How many miles can the car drive on 26 litres of petrol?

_____ **[2 marks]**

FM 2 Nutty Flake cereal is sold in two sizes.

The handy size contains 600 g and costs £1.55.

The large size contains 800 g and costs £2.20.

Which size is the better value?

_____ **[3 marks]**

3 A block of metal has a volume of 750 cm³ and a mass of 5.1 kg.

Calculate the density of the metal.

State the units of your answer.

_____ **[3 marks]**

PS 4 Josh cycles 20 km up a very steep hill at 10 km/h.

He turns round at the top and cycles back at 40 km/h.

His wife says that his average speed must be $\frac{10 + 40}{2} = 25$ km/h.

Is Josh's wife correct? Justify your answer.

_____

_____ **[3 marks]**

D

1 a Write down the value of:

  i $\sqrt{169}$

_____ [1 mark]

  ii 5^3

_____ [1 mark]

  b Write down the value of:

  i $\sqrt[3]{64}$

_____ [1 mark]

  ii 2^8

_____ [1 mark]

D

2 a Fill in the missing numbers.

			last digit
4^1	=	4	4
4^2	=	16	6
4^3	=	___	___
4^4	=	___	___
4^5	=	___	___

[2 marks]

  b What is the last digit of 4^{99}?

  Explain your answer.

_____ [1 mark]

  c Which is greater: 5^6 or 6^5?

  Justify your answer.

_____ [1 mark]

B

3 Write each of these as a fraction in its simplest form.

  a 8^{-1}

_____ [1 mark]

  b 4^{-2}

_____ [1 mark]

  c $2^{-3} + 4^{-1}$

_____ [2 marks]

B

PS 4 Arrange these numbers in order, starting with the smallest.

  8^{-1} 3^{-2} 2^{-4}

_____ _____ _____ [2 marks]

1 a Write $x^5 \times x^2$ as a single power of x.

_____ [1 mark]

b Write $x^8 \div x^4$ as a single power of x.

_____ [1 mark]

c i If $3^n = 81$, what is the value of n?

_____ [1 mark]

ii If $3^m = 27$, what is the value of m?

_____ [1 mark]

d Write down the value of $2^2 \times 5^2 \times 2^4 \times 5^4$.

_____ [1 mark]

2 Simplify the following expressions.

a $\dfrac{8a^3b \times 4a^2b^4}{2ab^2}$

_____ [2 marks]

b $(3x^2y^3)^2$

_____ [2 marks]

3 Write down the value of each of these.

a $8^{-\frac{4}{3}}$

_____ [1 mark]

b $121^{\frac{1}{2}}$

_____ [1 mark]

c $81^{-\frac{3}{4}}$

_____ [2 marks]

PS 4 Arrange these numbers in order, starting with the smallest.

$27^{-\frac{2}{3}}$ 2^{-3} $(\sqrt{16})^{-1}$

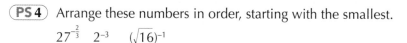

_____ _____ _____ [2 marks]

B

1 a Write the number 45.2×10^3 in standard form.

_____ **[1 mark]**

   b Write the number 0.6×10^{-2} as an ordinary number.

_____ **[1 mark]**

Here are six numbers.

2.5×10^5 1.8×10^6 45.2×10^3 8×10^{-4} 4.8×10^{-1} 0.6×10^{-2}

   c Which of the six numbers is the largest?

_____ **[1 mark]**

   d Which of the six numbers is the smallest?

_____ **[1 mark]**

   e Work out $2.5 \times 10^5 \times 8 \times 10^{-4}$.

Give your answer in standard form.   _____ **[1 mark]**

   f Work out $(4.8 \times 10^{-1}) \div (8 \times 10^{-4})$.

Give your answer in standard form.   _____ **[1 mark]**

C

2 a $\frac{1}{9} = 0.1111\ldots$ $\frac{2}{9} = 0.2222\ldots$

Use this information to write down the decimal equivalent of:

   i $\frac{4}{9}$ _____ **[1 mark]**

   ii $\frac{5}{9}$ _____ **[1 mark]**

   b Write down the reciprocal of each number.

Give your answers as fractions or mixed numbers as appropriate.

   i 10   _____ **[1 mark]**

   ii $\frac{3}{4}$ _____ **[1 mark]**

   c Write your answers to **b** as terminating or recurring decimals, as appropriate.

   i _____ **[1 mark]**

   ii _____ **[1 mark]**

   d Work out the reciprocal of each number.

   i 1.25   _____ **[1 mark]**

   ii 2.5   _____ **[1 mark]**

   iii 5   _____ **[1 mark]**

C

AU 3 The nth term of a series is $\dfrac{3n}{n + 1}$.

The first two terms are $\frac{3}{2} = 1.5$ and $\frac{6}{3} = 2$.

Show that the 6th term is the first term of the sequence that is not a terminating decimal.

_____ **[3 marks]**

1 a Simplify $\sqrt{48}$ as far as possible by writing it in the form $a\sqrt{b}$, where a and b are integers.

_____ **[2 marks]**

b Write $\sqrt{48} + \sqrt{75}$ in the form $p\sqrt{q}$, where p and q are integers.

_____ **[2 marks]**

c Expand and simplify $(\sqrt{3} + 5)(\sqrt{3} - 1)$.

_____ **[2 marks]**

d Expand and simplify $(\sqrt{5} + 2)(\sqrt{5} - 2)$.

_____ **[2 marks]**

2 a Rationalise the denominator of $\dfrac{3}{\sqrt{6}}$.

Simplify your answer as much as possible.

_____ **[2 mark]**

b This rectangle has an area of 20 cm².

Find the value of x.

Give your answer in surd form.

x cm

$5\sqrt{2}$

_____ cm **[2 marks]**

c Show clearly that $= \dfrac{4}{\sqrt{12}} + \dfrac{\sqrt{12}}{4} = \dfrac{7\sqrt{3}}{6}$.

_____ **[2 marks]**

AU 3 If $\sqrt{\dfrac{a}{b}} \times \sqrt{\dfrac{c}{d}} = N$, where N is a positive whole number, what relationship must exist between a, b, c and d?

_____ **[1 mark]**

1 The mass of a cube is directly proportional to the cube of its side.

Let m represent the mass of the cube.

Let s represent the side of the cube.

a Write down the proportionality equation that connects s and m.

_____ **[1 mark]**

b When $s = 10$, $m = 50$.

Find the value of the constant of proportionality, k.

_____ **[2 marks]**

c Find the value of m when $s = 20$.

_____ **[2 marks]**

d Find the value of s when $m = 3.2$.

_____ **[2 marks]**

2 a Two variables, a and b, are known to be proportional to each other.

When $a = 2$, $b = 4$.

Find the constant of proportionality, k, if:

i $a \propto b^2$

_____ **[2 marks]**

ii $a \propto \dfrac{1}{\sqrt{b}}$

_____ **[2 marks]**

b y is inversely proportional to the cube root of x.

When $y = 10$, $x = 8$.

Find the value of y when $x = 125$.

_____ **[3 marks]**

(PS 3) y is inversely proportional to the square root of x.

z is directly proportional to the cube of y.

When $x = 4$, $y = 3$ and $z = 9$.

Find the value of z when $x = \frac{1}{4}$.

_____ **[5 marks]**

1 a A coach is carrying 50 people, rounded to the nearest 10.

    i What is the smallest number of people that could have been on the coach?

        _____ **[1 mark]**

    ii What is the greatest number of people that could have been on the coach?

        _____ **[1 mark]**

b The coach is travelling at 50 mph, to the nearest 10 mph.

    i What is the lowest speed the coach could have been doing?

        _____ mph   **[1 mark]**

    ii What is the greatest speed the coach could have been doing?

        _____ mph   **[1 mark]**

2 A cube has a sides of 20 cm, measured to the nearest centimetre.

    a What is the lowest possible value of the volume of the cube?

        _____ cm³   **[2 marks]**

    b What is the greatest possible value of the volume of the cube?

        _____ cm³   **[2 marks]**

3 x and y are continuous values, both measured to 2 significant figures.

$x = 230$ and $y = 400$.

Work out the greatest possible value of $\dfrac{x}{y^2}$.

        _____ **[3 marks]**

PS 4 A sphere has a volume of 400 cm³, measured to the nearest 10 cm³.

What are the limits of the surface area of the sphere?

You will need to know that the volume of a sphere of radius r is $\frac{4}{3}\pi r^3$ and the surface area of a sphere of radius r is $4\pi r^2$.

        _____ **[4 marks]**

Number record sheet

Name _____   Marks _____

Form _____   Percentage _____

Date _____   Grade _____

Page number	Question number	Topic	Mark	Comments
124	1	Number	/4	
	2	Number	/4	
	3	Number	/3	
125	1	Number	/6	
	2	Number	/8	
	3	Number	/3	
126	1	Number	/7	
	2	Number	/6	
	3	Number	/2	
127	1	Fractions	/8	
	2	Fractions	/4	
	3	Fractions	/3	
128	1	Percentage	/3	
	2	Percentage	/4	
	3	Percentage	/3	
	4	Percentage	/2	
129	1	Percentage	/7	
	2	Percentage	/6	
	3	Percentage	/3	
130	1	Ratio	/4	
	2	Ratio	/2	
	3	Ratio	/6	
	4	Ratio	/3	
131	1	Ratio	/4	
	2	Ratio	/3	
	3	Ratio	/3	
	4	Ratio	/3	
132	1	Powers and reciprocals	/4	
	2	Powers and reciprocals	/4	
	3	Powers and reciprocals	/4	
	4	Powers and reciprocals	/2	
133	1	Powers and reciprocals	/5	
	2	Powers and reciprocals	/4	
	3	Powers and reciprocals	/4	
	4	Powers and reciprocals	/2	
134	1	Powers and reciprocals	/6	
	2	Powers and reciprocals	/9	
	3	Powers and reciprocals	/3	
135	1	Surds	/8	
	2	Surds	/6	
	3	Surds	/1	
136	1	Variation	/7	
	2	Variation	/7	
	3	Variation	/5	
137	1	Limits	/4	
	2	Limits	/4	
	3	Limits	/3	
	4	Limits	/4	
Total			/210	

Successes

1 _____

2 _____

3 _____

Areas for improvement

1 _____

2 _____

3 _____

1 a Work out the area of a circle of radius 12 cm.

Give your answer to 1 decimal place.

D

_____ cm^2 **[2 marks]**

b Work out the circumference of a circle of diameter 20 cm.

Give your answer to 1 decimal place.

_____ cm **[2 marks]**

2 Work out the area of a semicircle of diameter 20 cm.

Give your answer in terms of π.

C

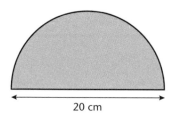

20 cm

_____ cm^2 **[2 marks]**

3 Calculate the area of this trapezium.

D

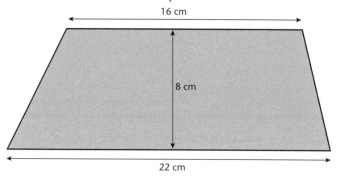

16 cm

8 cm

22 cm

_____ cm^2 **[2 marks]**

PS 4 a This square has the same numerical
value for its perimeter and its area.

x ☐

B

What is the value of x? _____ **[2 marks]**

b This circle has the same numerical
value for its circumference and its area.

r

What is the value of r? _____ **[2 marks]**

A

1 OAB is a minor sector of a circle of radius 10 cm.

Angle AOB $= 72°$.

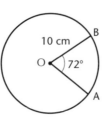

a Calculate the area of the **minor** sector OAB.

Give your answer in terms of π.

_____ cm² **[2 marks]**

b Calculate the perimeter of the **major** sector OAB.

Give your answer in terms of π.

_____ cm **[2 marks]**

D

2 A triangular prism has dimensions as shown.

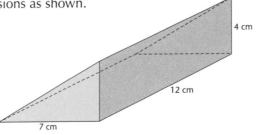

a Calculate the cross-sectional area of the prism.

_____ cm² **[2 marks]**

b Calculate the volume of the prism.

_____ cm³ **[2 marks]**

A

AU 3 The angle, θ, of a sector of a circle with radius r, and such that the arc length is equal to r, is called a **radian**.

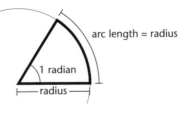

arc length = radius

1 radian

radius

Calculate the value, in degrees, of 1 radian.

Give your answer to 1 decimal place.

_____ **[3 marks]**

Cylinders and pyramids

1 A cylinder has a radius 4 cm and a height of 10 cm.

   a What is the volume of the cylinder?

     Give your answer in terms of π.

            _____ cm³ **[2 marks]**

   b What is the **total** surface area of the cylinder?

     Give your answer in terms of π.

            _____ cm² **[2 marks]**

B

2 The pyramid in the diagram has its top 3 cm cut off, as shown.
The shape which is left is called a frustum.
Calculate the volume of the frustum.

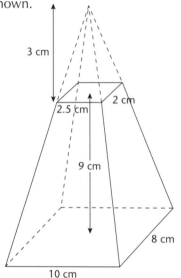

3 cm

2.5 cm   2 cm

9 cm

8 cm

10 cm

A*

            _____ cm³ **[3 marks]**

PS 3 A square-based pyramid just fits inside a cylinder of radius r and height h.

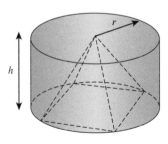

r

h

A

Show that the pyramid occupies approximately 21% of the volume of the cylinder.

            _____ **[5 marks]**

A

1 A fishing float is made of light wood with a density of 1.5 g/cm³.

The float is shaped as two cones, each of diameter 1 cm, joined at their bases.

One cone is 6 cm long and the other is 3 cm long.

Find the total mass of the float.

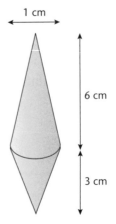

_____ g **[5 marks]**

A

2 A marble ashtray is made from a cylinder from which a hemisphere has been cut.

The cylinder has a radius of 6 cm and a depth of 6 cm.

The hemisphere has a radius of 5 cm.

What is the volume of the ashtray?

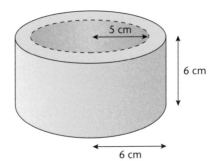

_____ cm³ **[4 marks]**

A* (PS 3) A sphere of **radius** r and a cone of **diameter** r and height h have the same volume.

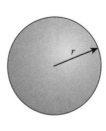
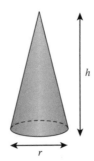

Find h in terms of r.

_____ **[3 marks]**

1 Calculate the length of the side marked x in this right-angled triangle.

Give your answer to 1 decimal place.

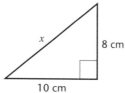

_____ cm **[3 marks]**

2 Calculate the length of the side marked x in this right-angled triangle.

Give your answer to 1 decimal place.

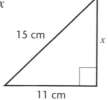

_____ cm **[3 marks]**

3 A flagpole 4 m tall is supported by a wire that is fixed at a point 2.1 m from the base of the pole.

How long is the wire? (The length is marked x on the diagram.)

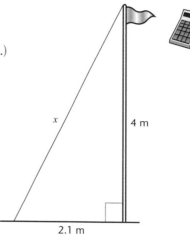

_____ m **[3 marks]**

PS 4 A series of right-angled triangles with one short side of length 1 cm are built onto successive hypotenuses, starting with the right-angled isosceles triangle with side 1 cm, as shown in the diagram.

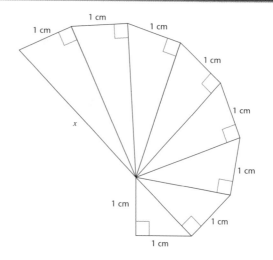

Calculate the length x.

_____ **[4 marks]**

1 Calculate the area of an isosceles triangle with sides of 12 cm, 12 cm and 7 cm.

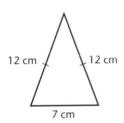

_____ cm² **[4 marks]**

2 The diagram shows a square-based pyramid with base length 10 cm and sloping edges 15 cm.

M is the mid-point of side AB, X is the mid-point of the base and E is directly above X.

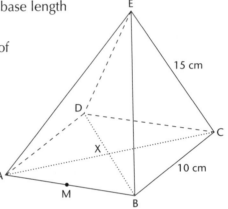

a Calculate the length of the diagonal AC.

_____ cm **[2 marks]**

b Calculate EX, the height of the pyramid.

_____ cm **[2 marks]**

c Using triangle ABE, calculate the length EM.

_____ cm **[2 marks]**

PS 3 A shed roof has a rectangular base 6 m by 3 m.

Each end is an equilateral triangle of side 3 m.

The ridge of the roof is 4 m long.

a Calculate the vertical height of the ridge of the roof (the distance from one of the ridge ends (A) to a point vertically below it (X).

Hint: Calculate the distance AM.

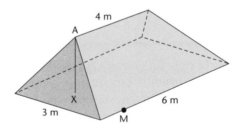

_____ m **[4 marks]**

b Calculate the volume of the roof.

Hint: Split the roof into three and put the two ends together to make a pyramid, leaving a triangular prism and a pyramid.

_____ m³ **[4 marks]**

Using Pythagoras and trigonometry

1 Use your calculator to work out sin (tan⁻¹ 0.65).

a Write down all of the numbers in the calculator display. _____ **[1 mark]**

b Round your answer to an appropriate degree of accuracy. _____ **[1 mark]**

2 Sine is defined as sin $a = \dfrac{\text{opposite}}{\text{hypotenuse}}$.

a Find the length of the side marked x in this right-angled triangle.

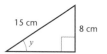

13 cm x 40°

_____ cm **[2 marks]**

b Find the size of the angle marked y in this right-angled triangle.

15 cm   8 cm y

_____ ° **[2 marks]**

3 Cosine is defined as cos $a = \dfrac{\text{adjacent}}{\text{hypotenuse}}$.

a Find the length of the side marked x in this right-angled triangle.

x 48°   15 cm

_____ cm **[2 marks]**

b Find the size of the angle marked y in this right-angled triangle.

20 cm y 9 cm

_____ ° **[2 marks]**

4 Tangent is defined as tan $a = \dfrac{\text{opposite}}{\text{adjacent}}$.

a Find the length of the side marked y in this right-angled triangle.

20° y 10 cm

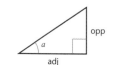

_____ cm **[2 marks]**

b Find the size of the angle marked x in this right-angled triangle.

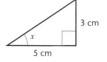

3 cm x 5 cm

_____ ° **[2 marks]**

AU 5 Use this triangle to explain the following rules.

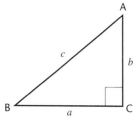

A c b B a C

a sin A = cos B

_____ **[2 marks]**

b tan A = $\dfrac{1}{\tan B}$

_____ **[2 marks]**

1 Find the size of the angle marked x in this right-angled triangle.

_____ ° **[3 marks]**

2 Find the length of the side marked x in this right-angled triangle.

_____ cm **[3 marks]**

3 Find the length of the side marked x in this right-angled triangle.

_____ cm **[3 marks]**

4 This diagram shows a track railway. It ascends to a height of 40 m using 50 m of track.

What angle does the track make with the horizontal?

_____ ° **[3 marks]**

PS 5 The first diagram shows a cross-section of a shed.

The first slope, from A to D, is 2 m long, at an angle 60° to the horizontal.

The second slope, from D to C, is 1.5 m long, at an angle 20° to the horizontal.

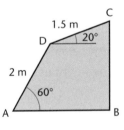

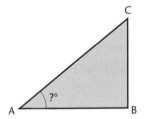

The roof gets damaged in a storm.

It is decided to replace the roof with a single slope from A to C, as shown in the second diagram.

What angle will this slope make with the horizontal?

_____ **[5 marks]**

1 ACD is a right-angled triangle.

B is a point on AC.

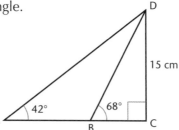

15 cm

42°   68°

A   B   C   D

a Calculate the length BC _____ cm **[3 marks]**

b Calculate the length AB _____ cm **[3 marks]**

2 O is the centre of a circle of radius 8 cm.

AB and BC are chords of equal length that intersect at B.

AO and OC intersect at 130°.

a Use the isosceles triangle OAC to work out the
length AC.

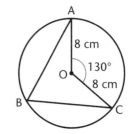

A

8 cm

130°

O   8 cm

B   C

_____ cm **[3 marks]**

b Use circle theorems to write down the size of angle ABC.

_____ ° **[1 mark]**

c Use the isosceles triangle ABC to work out the length BC.

_____ cm **[3 marks]**

PS 3 Calculate the angle CBA.

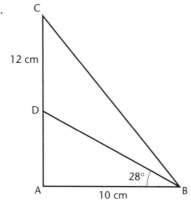

C

12 cm

D

28°

A   10 cm   B

_____ ° **[5 marks]**

1   A cuboid has sides of 6 cm, 8 cm and 15 cm.

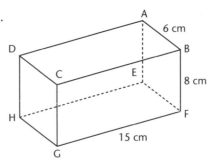

Calculate the size of angle BDF.

_____ ° **[4 marks]**

2   A triangular-based pyramid, ABCD, has an isosceles triangular base ABC, with AC = BC = 12 cm.

One face, ADB, is an equilateral triangle with sides of 8 cm.

D is directly above the mid-point, M, of the side AB.

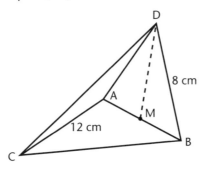

Find the size of angle DCM.

_____ ° **[5 marks]**

PS 3   A prism has a trapezoidal cross-section.

EH = 8 cm. AE = 5 cm. GF = 5 cm.

Angle EHD = Angle HGF = 90°.
Angle GAC = 30°.

Calculate the angle FAB.

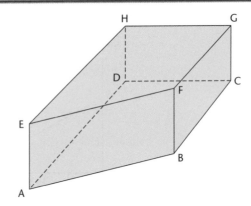

_____ ° **[6 marks]**

Trigonometric ratios of angles from 0° to 360°

1 The graph shows
$y = \sin x$ for
$0° \leqslant x \leqslant 360°$.

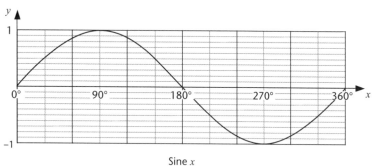

Sine x

a Use the graph to estimate two angles that have a sine of –0.6.

_____° and _____° **[2 marks]**

b You are told that $\sin^{-1} 0.15 = 8.6°$.

Write down two angles that have a sine of –0.15.

_____° and _____° **[2 marks]**

2 The graph shows
$y = \cos x$ for
$0° \leqslant x \leqslant 360°$.

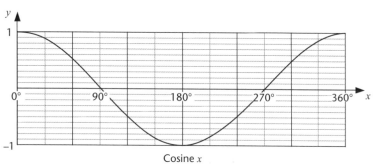

Cosine x

a Use the graph to estimate two angles that have a cosine of 0.75.

_____° and _____° **[2 marks]**

b You are told that $\cos^{-1} 0.25 = 75.5°$.

Write down two angles that have a cosine of –0.25.

_____° and _____° **[2 marks]**

AU 3 You are given that $\sin 38° = 0.616$ to 3 decimal places.

a Write down the sines of three other angles between 0° and 360°.

_____ **[2 marks]**

b Write down the cosines of four angles between 0° and 360°.

_____ **[2 marks]**

1   Find the size of angle ABC in this triangle.

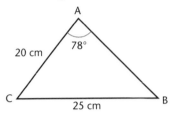

_____ ° **[3 marks]**

2 a   Find the value of x in this triangle.

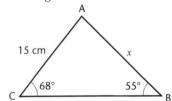

_____ cm **[3 marks]**

b   In this triangle, angle ABC is obtuse.

Find the size of angle ABC.

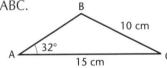

_____ ° **[3 marks]**

AU 3   Use the sine rule to find the sine of angle x.

Explain the result.

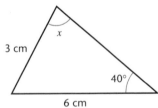

_____

_____ **[3 marks]**

1 Find the size of angle ACB in this triangle.

_____ ° **[3 marks]**

2 a Find the value of x in this triangle.

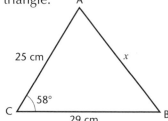

_____ cm **[3 marks]**

b In this triangle, angle ABC is obtuse.

Find the size of angle ABC.

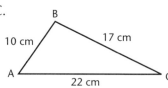

_____ ° **[3 marks]**

AU 3 Use the cosine rule to find the cosine of angle x.

Explain the result.

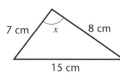

_____

_____ **[3 marks]**

A

1 ABCD is a quadrilateral.

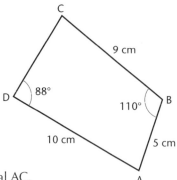

a Find the length of the diagonal AC.

_____ cm **[3 marks]**

b Find the size of angle DCA.

_____ ° **[3 marks]**

A

2 a Work out the area of this triangle.

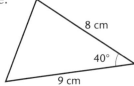

_____ cm² **[2 marks]**

b In the triangle PQR, PQ = 7 cm, PR = 12 cm and angle PQR = 106°.

Work out the area of PQR.

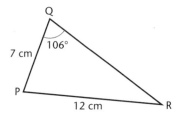

_____ cm² **[4 marks]**

A*

PS 3 The area of this quadrilateral is 60 cm².

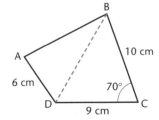

Calculate the length AB. _____ **[9 marks]**

PS 1 The diagram shows a kite, ABCD, joined to a parallelogram, CDEF.

Angle BAD = 100° and angle CFE = 60°.

When the side of the kite is extended, it passes along the diagonal of the parallelogram shown by the dotted line.

Use the properties of quadrilaterals to find the size of angle CED. Show all your working.

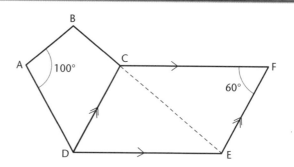

_____ ° **[4 marks]**

2 a Explain why the interior angles of a pentagon add up to 540°.

_____
_____
_____
_____
_____ **[2 marks]**

b The diagram shows three sides of a regular polygon.

The exterior angle is 36°.

How many sides does the polygon have altogether?

_____ **[2 marks]**

c The interior angle of a regular polygon is 160°.

Explain why the polygon must have 18 sides. _____ **[2 marks]**

AU 3 These are the standard quadrilaterals.

Square   Rectangle   Rhombus   Kite   Parallelogram   Trapezium

All quadrilaterals have four sides.

a State a mathematical property that the **square** and the **rectangle** have in common. _____ **[1 mark]**

b State a mathematical property that the **kite** and the **rhombus** have in common. _____ **[1 mark]**

c State a mathematical property that the **parallelogram** and the **rectangle** have in common. _____ **[1 mark]**

d Choose any other pair of quadrilaterals and write down a mathematical property that they have in common.

Quadrilateral 1 _____

Quadrilateral 2 _____

Property _____

[1 mark]

Circle theorems

B

1 In the diagram, O is the centre of the circle. ABC is a triangle, AB is a diameter of the circle and angle CBA = 63°.

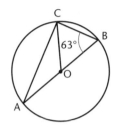

a Find the size of angle CAB.

_____ °

Give a reason for your answer.

_____

_____ **[2 marks]**

b Explain why the angle OCB is 63°.

_____

_____ **[1 mark]**

B

2 In the diagram, O is the centre of the circle.

a State the value of x.

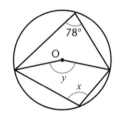

_____ °

Give a reason for your answer.

_____

_____ **[2 marks]**

b State the value of y.

_____ °

Give a reason for your answer.

_____

_____ **[2 marks]**

B

PS 3 Three points A, B and C are shown on a centimetre grid.

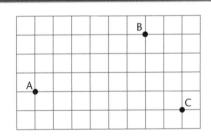

A circle is drawn through all three points.

a What is the radius of the circle?

Give your answer as an exact value. _____ **[2 marks]**

b Mark the centre of the circle on the grid.

_____ **[1 mark]**

AQA 3   EDEXCEL 2/3   OCR 2

1   In the diagram, PT is a tangent to the circle, centre O, at Q.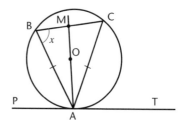
AB is a diameter of the circle and angle QBA = 57°.

a   Calculate the size of angle QAB.

Give a reason for your answer.   ∠QAB = _____ °

_____

_____ **[2 marks]**

b   Calculate the size of angle BQT.

Give reasons for your answers.   ∠BQT = _____ °

_____

_____ **[2 marks]**

AU 2   The diagram shows a circle with centre O. ABC is an
isosceles triangle with AB = AC.

PT is a tangent to the circle at A, M is the mid-point
of BC and angle ABC = $x°$.

Explain why AM must pass through the centre of
the circle.

Give reasons for any values or angles you write down.

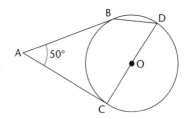

_____

_____

_____

_____ **[3 marks]**

PS 3   AB and AC are tangents to a circle centre O.

D is a point on the circumference of the circle such that CD is a diameter.

The angle BAC is 50°.

Calculate the value of the
angle BDO.

Give reasons for any angles
you write down or calculate.

_____ ° **[4 marks]**

Transformation geometry

1 Triangles A, P, Q, R, S and T are not drawn accurately.

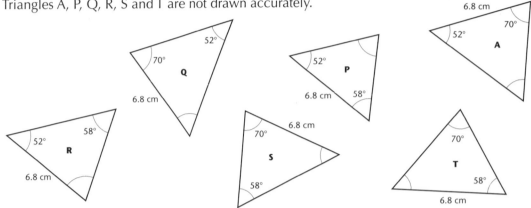

Which two of triangles P, Q, R, S and T are congruent to triangle A?

_____ and _____ **[2 marks]**

2 a Describe the transformation that takes the shaded triangle to triangle A.

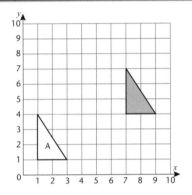

_____ **[2 marks]**

b Translate the shaded triangle by $\binom{-4}{2}$.

Label the image B. **[1 mark]**

c The shaded triangle is translated by $\binom{7}{-5}$ to give triangle C.

What vector will translate triangle C to the shaded triangle?

_____ **[1 mark]**

AU 3 Bashir and Mona are constructing triangles.

My triangle has one side 7 cm long, one side 8 cm long and one angle of 40°.

My triangle also has one side 7 cm long, one side 8 cm long and one angle of 40°, so it must be congruent to yours.

Sketch two triangles, marking on the lengths and angles to show that Mona is wrong.

[3 marks]

1 **a** Reflect the shaded triangle in the *y*-axis.
Label the image A.   [1 mark]

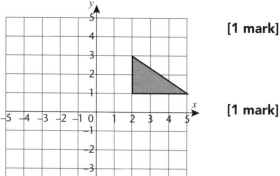

b Reflect the shaded triangle in the line
$y = -1$.
Label the image B.   [1 mark]

c Reflect the shaded triangle in the line
$y = -x$.
Label the image C.   [1 mark]

D

2 **a** Rotate the shaded triangle by 90°
clockwise about (1, 0).
Label the image A.   [1 mark]

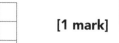

b Rotate the shaded triangle by a half-turn
about (1, 3).
Label the image B.   [1 mark]

c What rotation will take triangle A to
triangle B?   [3 marks]

D

PS 3 Square A can be transformed to square B by a translation, a reflection and a rotation.

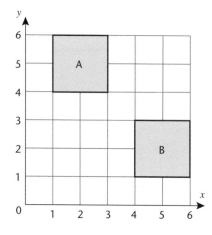

B

Describe each of these.

Translation of $\begin{pmatrix} ...... \\ ...... \end{pmatrix}$ [1 mark]

Reflection in _____   [1 mark]

Rotation of _____   [2 marks]

C **1**

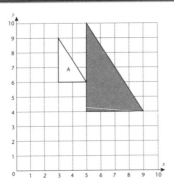

a What transformation takes the shaded triangle to triangle A?

_____ **[2 marks]**

b Draw the image after the shaded triangle is enlarged
by a scale factor $\frac{1}{4}$ about (1, 0). **[2 marks]**

C **2** **a** Rotate the shaded triangle 90°
clockwise about the origin.
Label the image A. **[1 mark]**

b Reflect triangle A in the y-axis.
Label the image B. **[1 mark]**

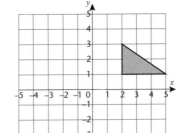

c What **single** transformation maps the shaded triangle to triangle B?

_____ **[1 mark]**

A **PS 3** Triangle A is an enlargement of the shaded
triangle, scale factor $\frac{1}{3}$, centre (1, 9).

Triangle B is an enlargement of triangle A,
scale factor 2, centre (0, 12).

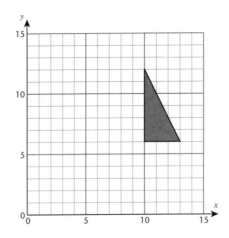

What transformation takes triangle B to the shaded triangle?

_____ **[3 marks]**

1 Make an accurate drawing of this triangle.

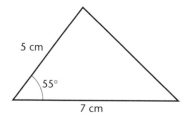

[3 marks]

2 a Using compasses and a ruler, construct an angle of 60° at the point A.

A •————————————

[2 marks]

b i Use compasses and a ruler to construct this triangle accurately.

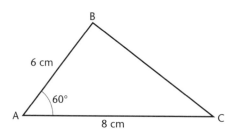

[3 marks]

ii Measure the length of the line BC. _____ cm **[1 mark]**

AU 3 Describe how to construct an angle of 30° using a straight edge and a ruler.

_____ **[2 marks]**

1 Use compasses and a ruler to do these constructions.

   a Construct the perpendicular bisector of AB.

A●

●B

[2 marks]

   b Construct the perpendicular
      at the point C to the line L.

————————————●————————————L
             C

[2 marks]

2 Use compasses and ruler to do these constructions.

   a Construct the perpendicular bisector of the line L.

————————————————— L

[2 marks]

   b Construct the bisector of angle ABC.

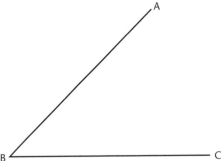

[2 marks]

AU 3 Describe how to divide a line exactly into four equal pieces using a straight edge and
a pair of compasses.

_____ **[2 marks]**

1 Use compasses and ruler to construct
the perpendicular from the point C
to the line L.

• C

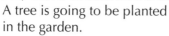

L

[3 marks]

2 The diagram, which is drawn to scale, shows a flat, rectangular lawn of length 10 m and
width 6 m, with a circular flower bed of radius 2 m.

A tree is going to be planted
in the garden.

It has to be at least 1 metre
from the edge of the garden
and at least 2 metres from the
flower bed.

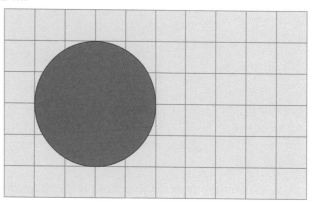

a Draw a circle to show the area around the flower bed where the tree *cannot*
be planted.

[1 mark]

b Show the area of the garden in which the tree can be planted.

[2 marks]

AU 3 The inscribed circle of a triangle is the circle with a
centre at the intersection of the angle bisector of
each vertex.

The circle just touches each side of the triangle.

Three triangles, A, B and C, are formed by the sides
of the triangle and the angle bisectors.

Explain why the sides a, b and c are in the same
ratio as the areas of the triangles A, B and C,

i.e. $a : b : c$ = Area A : Area B : Area C.

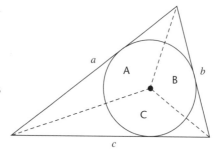

[3 marks]

B

1 Triangle ABC is similar to triangle CDE.

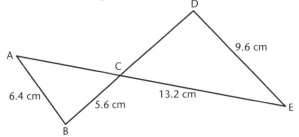

a Find the length of CD. _____ cm [2 marks]

b Find the length of AC _____ cm [1 mark]

A

2 A pinhole camera projects an image through a pinhole onto light-sensitive paper.
The paper is 12 cm by 8 cm and the camera box is 20 cm deep.

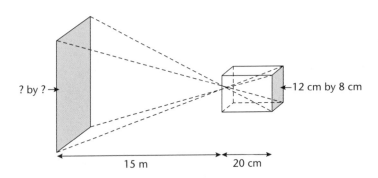

When the subject of the photo is 15 metres from the front of the camera, what are the
width and depth of the total area that will be photographed?

width _____ m depth _____ m [3 marks]

A

PS 3 Archie has an eyeline that is 1.2 m off the ground.

He is standing 5 m from his father and sister.

His father, who is 1.8 m tall,
and his sister, who is 80 cm
tall, stand next to each other.

Archie notices that he can
just see the base of a tree
over his sister's head, and
he can just see the top of the
tree over his father's head.

How tall is the tree?

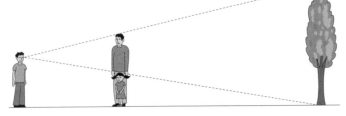

_____ [4 marks]

1 In a 'home' snooker set, everything is made to three-quarters of the size of regular snooker equipment.

  a A full-size snooker table is 3.6 m by 1.8 m.

    What is the size of a 'home' snooker table?

    _____ m by _____ m **[2 marks]**

  b The manufacturers recommend a room with an area of 52 m² for a regular snooker table.

    What room area should be recommended for a 'home' snooker set?

    _____ m² **[2 marks]**

  c Regular snooker balls have a volume of 75 cm³. What is the volume of a 'home' snooker ball?

    _____ cm³ **[2 marks]**

2 Ecowash washing powder is sold in two sizes: standard and family.

It is sold in cuboidal boxes that are similar in shape.

The family size contains 2.5 kg of powder.

The standard size contains 1 kg of powder.

The amount of powder is proportional to the volume of the boxes.

  a The height of the standard-size box is 21 cm.

    What is the height of the family-size box? _____ **[2 marks]**

  b The labels on the front of each box are also similar in shape.

    The area of the label on the family-size box is 70 cm².

    What is the area of the label on the standard-size box? _____ **[2 marks]**

PS 3 When the toy alien in this packet is placed in water it grows in size from 5 cm high to 9 cm high.

Is the claim on the packet justified?

Place me in water and watch me grow to 6 times bigger!

MAGIC ALIEN

Not suitable for children under 8 years old

_____ **[3 marks]**

1 On this grid $\overrightarrow{OA} = \mathbf{a}$ and $\overrightarrow{OB} = \mathbf{b}$.

a Label each of the following points on the grid.

i C such that:

$\overrightarrow{OC} = 2\mathbf{b}$

ii D such that:

$\overrightarrow{OD} = 2\mathbf{a} + \mathbf{b}$

iii E such that:

$\overrightarrow{OE} = -\mathbf{a}$

iv F such that:

$\overrightarrow{OF} = 2\mathbf{b} - 2\mathbf{a}$

[4 marks]

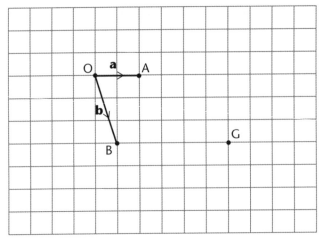

b Give $\overrightarrow{OG}$ in terms of $\mathbf{a}$ and $\mathbf{b}$. _____ **[1 mark]**

c The point H, which is not on the diagram, is such that

$\overrightarrow{OH} = -5\mathbf{a} + 7\mathbf{b}$.

Write down the vector $\overrightarrow{HO}$. _____ **[1 mark]**

(AU 2) $\mathbf{p}_1$ and $\mathbf{p}_2$ are perpendicular vectors.

$\mathbf{p}_1$ is the vector $\mathbf{a} + 3\mathbf{b}$, which can be written as $\begin{pmatrix} 1 \\ 3 \end{pmatrix}$.

$\mathbf{p}_2$ is the vector $3\mathbf{a} - \mathbf{b}$, which can be written as $\begin{pmatrix} 3 \\ -1 \end{pmatrix}$.

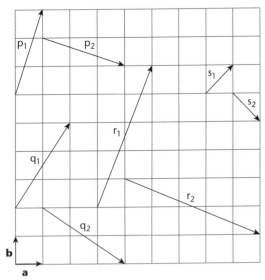

a Write the pairs of vectors $\mathbf{q}_1$, $\mathbf{q}_2$; $\mathbf{r}_1$, $\mathbf{r}_2$; $\mathbf{s}_1$, $\mathbf{s}_2$ in the same way.

$\mathbf{q}_1 \begin{pmatrix} \end{pmatrix}$ $\qquad$ $\mathbf{q}_2 \begin{pmatrix} \end{pmatrix}$

$\mathbf{r}_1 \begin{pmatrix} \end{pmatrix}$ $\qquad$ $\mathbf{r}_2 \begin{pmatrix} \end{pmatrix}$

$\mathbf{s}_1 \begin{pmatrix} \end{pmatrix}$ $\qquad$ $\mathbf{s}_2 \begin{pmatrix} \end{pmatrix}$

[3 marks]

b If $\begin{pmatrix} a \\ b \end{pmatrix}$ and $\begin{pmatrix} c \\ d \end{pmatrix}$ are perpendicular vectors, what relationship exists between a, b, c and d?

_____ **[1 mark]**

1   OABC is a quadrilateral.

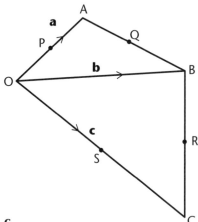

$\vec{OA}$ = **a**, $\vec{OB}$ = **b** and $\vec{OC}$ = **c**.

The mid-points of OA, AB, BC and OC are P, Q, R and S respectively.

a   Give the following vectors in terms of **a**, **b** and **c**.

   i $\vec{PS}$ _____ **[1 mark]**

   ii $\vec{AB}$ _____ **[1 mark]**

   iii $\vec{BC}$ _____ **[1 mark]**

   iv $\vec{QR}$ _____ **[1 mark]**

b   What type of quadrilateral is PQRS?

   Explain your answer.

_____

_____

_____

_____ **[1 mark]**

PS 2   OACB is a parallelogram.

$\vec{OA}$ = **a** and $\vec{OB}$ = **b**

$\vec{OM} : \vec{MA}$ = 4 : 1

$\vec{ON} : \vec{NC}$ = 2 : 1

MNX is a straight line.

Work out the ratio BX : XC.

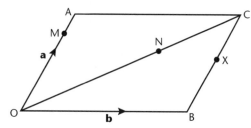

_____ **[6 marks]**

Geometry record sheet

Name _____   Marks _____
Form _____   Percentage _____
Date _____   Grade _____

Page number	Question number	Topic	Mark	Comments
139	1	Circles and area	/4	
	2	Circles and area	/2	
	3	Circles and area	/2	
	4	Circles and area	/4	
140	1	Sectors and prisms	/4	
	2	Sectors and prisms	/4	
	3	Sectors and prisms	/3	
141	1	Cylinders and pyramids	/4	
	2	Cylinders and pyramids	/3	
	3	Cylinders and pyramids	/5	
142	1	Cones and spheres	/5	
	2	Cones and spheres	/4	
	3	Cones and spheres	/3	
143	1	Pythagoras' theorem	/3	
	2	Pythagoras' theorem	/3	
	3	Pythagoras' theorem	/3	
	4	Pythagoras' theorem	/4	
144	1	Pythagoras' theorem	/4	
	2	Pythagoras' theorem	/6	
	3	Pythagoras' theorem	/8	
145	1	Using Pythagoras and trigonometry	/2	
	2	Using Pythagoras and trigonometry	/4	
	3	Using Pythagoras and trigonometry	/4	
	4	Using Pythagoras and trigonometry	/4	
	5	Using Pythagoras and trigonometry	/4	
146	1	Using Pythagoras and trigonometry	/3	
	2	Using Pythagoras and trigonometry	/3	
	3	Using Pythagoras and trigonometry	/3	
	4	Using Pythagoras and trigonometry	/3	
	5	Using Pythagoras and trigonometry	/5	
147	1	More advanced trigonometry	/6	
	2	More advanced trigonometry	/7	
	3	More advanced trigonometry	/5	
148	1	3-D trigonometry	/4	
	2	3-D trigonometry	/5	
	3	3-D trigonometry	/6	
149	1	Trigonometric ratios of angles from 0° to 360°	/4	
	2	Trigonometric ratios of angles from 0° to 360°	/4	
	3	Trigonometric ratios of angles from 0° to 360°	/4	
150	1	Sine rule	/3	
	2	Sine rule	/6	
	3	Sine rule	/3	
151	1	Cosine rule	/3	

Page number	Question number	Topic	Mark	Comments
	2	Cosine rule	/6	
	3	Cosine rule	/3	
152	1	Solving triangles	/6	
	2	Solving triangles	/6	
	3	Solving triangles	/9	
153	1	Polygons	/4	
	2	Polygons	/6	
	3	Polygons	/4	
154	1	Circle theorems	/3	
	2	Circle theorems	/4	
	3	Circle theorems	/3	
155	1	Circle theorems	/4	
	2	Circle theorems	/3	
	3	Circle theorems	/4	
156	1	Transformation geometry	/2	
	2	Transformation geometry	/4	
	3	Transformation geometry	/3	
157	1	Transformation geometry	/3	
	2	Transformation geometry	/5	
	3	Transformation geometry	/4	
158	1	Transformation geometry	/4	
	2	Transformation geometry	/3	
	3	Transformation geometry	/3	
159	1	Constructions	/3	
	2	Constructions	/6	
	3	Constructions	/2	
160	1	Constructions	/4	
	2	Constructions	/4	
	3	Constructions	/2	
161	1	Constructions and loci	/3	
	2	Constructions and loci	/3	
	3	Constructions and loci	/3	
162	1	Similarity	/3	
	2	Similarity	/3	
	3	Similarity	/4	
163	1	Similarity	/6	
	2	Similarity	/4	
	3	Similarity	/3	
164	1	Vectors	/6	
	2	Vectors	/4	
165	1	Vectors	/5	
	2	Vectors	/6	
Total			**/343**	

Successes

1 _____

2 _____

3 _____

Areas for improvement

1 _____

2 _____

3 _____

C

1 a Work out the value of $2a^2 - 3b$, when $a = 3$ and $b = -2$.

_____ **[2 marks]**

b Work out the value of $\dfrac{x^2 - y^2}{z}$, when $x = -2$, $y = -6$, and $z = -8$.

_____ **[2 marks]**

C

2 a Expand $5(x - 3)$.

_____ **[1 mark]**

b Expand and simplify $2(x + 1) + 2(3x + 2)$.

_____ **[2 marks]**

c Expand and simplify $3(x - 4) + 2(4x + 1)$.

_____ **[2 marks]**

d A rectangle has a length of $2x + 3$ and a width of $x + 3$.

Write down and simplify an expression for the perimeter in terms of x.

$$\begin{array}{c} \xleftarrow{\hspace{2cm} 2x + 3 \hspace{2cm}} \\ \left. \rule{0pt}{1cm} \right\updownarrow x + 3 \end{array}$$

_____ **[2 marks]**

C

3 Factorise these expressions.

a $4x + 6$

_____ **[1 mark]**

b $5x^2 + 2x$

_____ **[1 mark]**

D

AU 4 a and b are different positive whole numbers.

Choose values for a and b so that the formula

$7a - 3b$

a evaluates to a positive odd number.

_____ **[1 mark]**

b evaluates to a prime number.

_____ **[1 mark]**

c evaluates to a positive even number.

_____ **[1 mark]**

1 Solve these equations.

a $3x + 4 = 1$

$x =$ _____ **[2 marks]**

b $\dfrac{7x - 2}{3} = 4$

$x =$ _____ **[3 marks]**

c $4(x - 1) = 6$

$x =$ _____ **[3 marks]**

2 Solve these equations.

a $4(3y - 2) = 16$

$y =$ _____ **[3 marks]**

b $5x - 2 = x + 10$

$x =$ _____ **[3 marks]**

c $5x - 2 = 3x + 1$

$x =$ _____ **[3 marks]**

PS 3 A rectangle with sides 4 and $3x + 1$ has a smaller rectangle with sides 2 and $2x - 1$ cut from it.

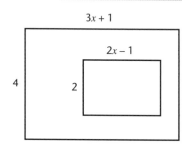

The area remaining is 16 cm².

Work out the value of x.

_____ **[3 marks]**

1 Solve these equations.

a $3(x + 4) = x - 5$

$x =$ _____ **[3 marks]**

b $5(x - 2) = 2(x + 4)$

$x =$ _____ **[3 marks]**

2 ABC is a triangle with sides, measured in centimetres, of x, $3x - 1$ and $2x + 5$.

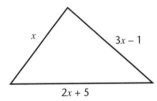

If the perimeter of the triangle is 25 cm, find the value of x.

_____ **[3 marks]**

PS 3 Will has eight packets of mints and eight loose mints.

Chas has 11 packets of mints and nine loose mints.

Will says to Chas 'If you give me one of your packets and all of your loose mints we will have the same total number of mints'.

How many mints are in a packet?

_____ **[3 marks]**

1 Use trial and improvement to solve the equation $x^3 + 4x = 203$.

The first two entries of the table are filled in.

Complete the table to find the solution. Give your answer to 1 decimal place.

Guess	$x^3 + 4x$	Comment
5	145	Too low
6	240	Too high

$x =$ _____ **[3 marks]**

2 Darlene is using trial and improvement to find a solution to this equation.

$2x + \dfrac{2}{x} = 8$

The table shows her first trial.

Complete the table to find the solution. Give your answer to 1 decimal place.

Guess	$2x + \frac{2}{x}$	Comment
3	6.66	Too low

$x =$ _____ **[4 marks]**

PS 3 A cuboidal juice carton holds 2 litres (2000 cm³).

The base is a square.

The height is 6 cm more than the base.

Use trial and improvement to find the side of the square base to one decimal place.

_____ **[4 marks]**

Simultaneous equations

AQA 2/3   EDEXCEL 3   OCR 2

B

1 Solve these simultaneous equations.

   a $5x + 3y = 8$

      $3x - y = 9$

$x =$ _____

$y =$ _____ **[3 marks]**

   b $6x + 3y = 9$

      $2x - 3y = 1$

$x =$ _____

$y =$ _____ **[3 marks]**

   c $5x + 3y = 6$

      $3x - 7y = 19$

$x =$ _____

$y =$ _____ **[4 marks]**

B

2 **a** Rearrange the equation $y + 3x = 5$ to make y the subject.

$y =$ _____ **[1 mark]**

   b Solve these simultaneous equations by the method of substitution.

      $y + 3x = 5$

      $2x = 22 - 3y$

$x =$ _____

$y =$ _____ **[3 marks]**

A

FM 3 Three sacks of cement and five pallets of bricks weigh 375 kg.

Two sacks of cement and one pallet of bricks weigh 110 kg.

Baz needs six sacks of cement and six pallets of bricks.

Will he be able to carry them in his trailer, which has a safe working load of half a tonne? Justify your answer.

_____ **[4 marks]**

Algebra

Simultaneous equations and formulae

AQA 2/3 EDEXCEL 2/3
OCR 1/3

FM 1 A widget weighs x grams. A whotsit weighs y grams.

24 widgets and 20 whotsits weigh 134 grams.

20 widgets and 24 whotsits weigh 130 grams.

a Write down a pair of simultaneous equations in x and y, using the information above.

_____ **[2 marks]**

b Solve your simultaneous equations and find the weight of a widget.

_____ **[3 marks]**

2 Rearrange each of these formulae to make x the subject.

Simplify your answers where possible.

a $C = \pi x$

$x =$ _____ **[1 mark]**

b $6y = 3x - 9$

$x =$ _____ **[2 marks]**

c $3q = 2(4 - x)$

$x =$ _____ **[2 marks]**

d $4y + 3 = 2x + 1$

$x =$ _____ **[2 marks]**

e $4(y + 3) = 2(x + 1)$

$x =$ _____ **[3 marks]**

PS 3 Keith notices that the cost of six mince pies is £1.24 less than the price of 10 gingerbread men.

Let the price of a mince pie be x pence and the price of a gingerbread man be y pence.

a Express the cost of one gingerbread man, y, in terms of the price of a mince pie, x.

_____ **[2 marks]**

b If the price of mince pie is 21p, how much is a gingerbread man?

_____ **[2 marks]**

C

1 a Expand $3(x + 2)$.

_____ [1 mark]

b Expand $x(x + 2)$.

_____ [1 mark]

c Expand and simplify $(x - 3)(x + 2)$.

_____ [2 marks]

d A rectangle has length $x + 2$ and width $x + 1$.

Write down an expression for the area, in terms of x, and simplify it.

_____ [2 marks]

C

2 a Multiply out and simplify $(x - 4)(x + 1)$.

_____ [2 marks]

b Multiply out and simplify $(x + 4)^2$.

_____ [2 marks]

E

3 a Work out the value of $3p + 2q$ when $p = -2$ and $q = 5$.

_____ [2 marks]

b Find the value of $a^2 + b^2$ when $a = 4$ and $b = 6$.

_____ [2 marks]

c An aeroplane has f first-class seats and e economy seats.

For a flight, each first-class seat costs £200 and each economy seat costs £50.

i If all seats are taken, write down an expression in terms of f and e for the total cost of all the seats in the aeroplane.

_____ [1 mark]

ii If $f = 20$ and $e = 120$, work out the actual cost of all the seats.

_____ [2 marks]

B

AU 4 Find values of a, b and c that make this identity true.

$x^2 + ax + 12 \equiv (x + b)(x + c)$

_____ [2 marks]

Factorising quadratic expressions

1 **a** Factorise $x^2 - 4x$.

_____ [1 mark]

   b **i** Factorise $x^2 - 4x - 12$.

_____ [2 marks]

   ii Solve the equation $x^2 - 4x - 12 = 0$.

_____ [1 mark]

2 **a** Expand and simplify $(x + y)(x - y)$.

_____ [1 mark]

   b Factorise $x^2 - 49$.

_____ [1 mark]

   c Factorise $4a^2 - 9$.

_____ [2 marks]

3 **a** Expand and simplify $(2x + 3)(3x - 1)$.

_____ [2 marks]

   b Factorise $6x^2 - 17x + 12$.

_____ [2 marks]

4 Solve these equations.

   a $x^2 + 3x - 10 = 0$

$x =$ _____

$x =$ _____ [3 marks]

   b $x^2 + 4x - 5 = 0$

$x =$ _____

$x =$ _____ [3 marks]

5 **a** Expand $(2x - 5)(2x + 5)$.

_____ [1 mark]

   b Use your answer to part **a** to write down a product of two numbers with a difference of 10, that make

   i 375

_____ [1 mark]

   ii 2475

_____ [1 mark]

Solving quadratic equations

A

1 a Factorise $12x^2 + 7x + 1$.

_____ **[2 marks]**

b Hence solve $12x^2 + 7x + 1 = 0$.

$x =$ _____

$x =$ _____ **[1 mark]**

A

2 Solve these equations.

a $2x^2 - 9 = 0$

$x =$ _____

$x =$ _____ **[2 marks]**

b $4x^2 - 20x = 0$

$x =$ _____

$x =$ _____ **[2 marks]**

A

3 a Solve the equation $x^2 + 3x - 7 = 0$. Give your answers to 2 decimal places.

$x =$ _____

$x =$ _____ **[3 marks]**

b Solve the equation $2x^2 + 5x - 4 = 0$. Give your answers to 2 decimal places.

$x =$ _____

$x =$ _____ **[3 marks]**

A

4 Solve these equations.

a $9x^2 - 9x + 2 = 0$

$x =$ _____

$x =$ _____ **[3 marks]**

b $8x^2 + 6x - 5 = 0$

$x =$ _____

$x =$ _____ **[3 marks]**

A

PS 5 A woman is x years old.

Her husband is six years younger than her.

The product of their ages is 1360.

How old is the woman?

_____ **[4 marks]**

1 a Find the values of a and b such that

$x^2 + 8x - 7 = (x + a)^2 - b$.

$a =$ _____

$b =$ _____ **[2 marks]**

b Hence solve $x^2 + 8x - 7 = 0$.

Give your answers in surd form.

$x =$ _____

$x =$ _____ **[2 marks]**

2 a Find the values of a and b such that

$x^2 + 10x - 3 = (x + a)^2 - b$.

$a =$ _____

$b =$ _____ **[2 marks]**

b Hence solve $x^2 + 10x - 3 = 0$.

Give your answers to 2 decimal places.

$x =$ _____

$x =$ _____ **[2 marks]**

3 a Find the values of a and b such that

$x^2 + 2x + 7 = (x + a)^2 + b$.

$a =$ _____

$b =$ _____ **[2 marks]**

b Explain why the equation $x^2 + 2x + 7 = 0$ has no solution.

_____ **[1 mark]**

4 a Find the values of a and b such that

$x^2 - 8x + 2 = (x + a)^2 - b$.

$a =$ _____

$b =$ _____ **[2 marks]**

b Hence solve $x^2 - 8x + 2 = 0$.

Give your answers to 2 decimal places.

$x =$ _____

$x =$ _____ **[2 marks]**

PS 5 Dave rewrites the expression $x^2 + px + q$ by completing the square.

He does this correctly, and gets $(x - 5)^2 - 15$.

What are the values of p and q?

_____ **[2 marks]**

PS 1 Use the quadratic formula to solve the equation $x^2 + 5x - 9 = 0$.

Leave your answers in surd form.

$x =$ _____ **[2 marks]**

PS 2 The sides of a right-angled triangle are $(5x + 4)$ cm, $(6x + 1)$ cm and $(2x - 1)$ cm.

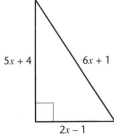

$5x + 4$ $6x + 1$

$2x - 1$

a Explain why $7x^2 - 24x - 16 = 0$.

_____

_____ **[2 marks]**

b Find the value of x.

_____ cm **[3 marks]**

3 A number plus its reciprocal equals 2.9.

Let the number be x.

a Set up a quadratic equation in x.

_____ **[2 marks]**

b Solve the equation to find x.

$x =$ _____ **[3 marks]**

FM 4 The width of a rectangular room is 2 metres less than the length. It cost £183.60 to carpet the room. Carpet costs £19.20 per square metre. How wide is the room?

_____ **[4 marks]**

D

FM 1 Martin walked from his house to a viewpoint 5 kilometres away, and back again.

The distance–time graph shows his journey.

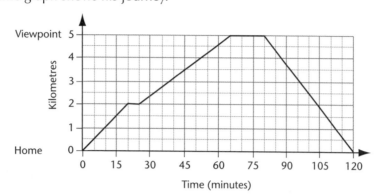

a The viewpoint is uphill from Martin's house.

Martin took a rest before walking up the steepest part of the hill.

  i How far from home was Martin when he took a rest?_____ km   **[1 mark]**

  ii For how long did Martin rest? _____ minutes   **[1 mark]**

b Martin stopped at the viewpoint before returning home.

He then walked home at a fast, steady pace.

  i How long did it take Martin to walk home? _____ minutes   **[1 mark]**

  ii What was Martin's average speed on the way home? _____ km/hour   **[2 marks]**

B

2 A water tank is empty. It holds a total of 10 000 litres. It is filled at a rate of 2000 litres per minute for 3 minutes and then 1000 litres per minute for 4 minutes.

It then empties at a rate of 2500 litres per minute.

Show this information on these axes.

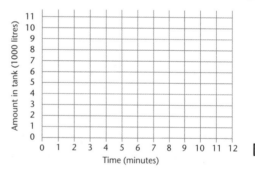

[3 marks]

C

PS 3 A jogger sets off at 10 am from point P to run along a trail at a steady pace of 12 km per hour.

60 minutes later, his wife sets off on a bicycle from P on the same trail at a steady pace of 20 km per hour. After 10 km she gets a puncture which takes 15 minutes to repair. She then sets off again at a steady pace of 20 km per hour.

The couple have left their car at a point 30 km along the trail.

a Represent these journeys on a graph with a horizontal time axis from 10 am to 1 pm and a vertical distance axis from 0 to 35 km.   **[2 marks]**

b Who arrives at the car first and by how long?

_____ **[1 mark]**

D

1 Draw the graph of $y = 2x - 1$ for $-3 \leqslant x \leqslant 3$ on the axes below.

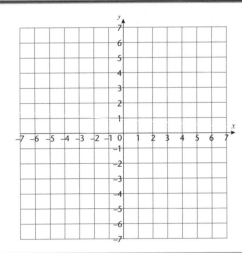

[2 marks]

C

2 a What is the gradient of each of these lines?

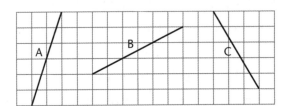

Line A: _____ Line B: _____ Line C: _____ [3 marks]

b On the grid below, draw lines with gradients of:

i 2 **ii** $-\frac{1}{2}$ **iii** $\frac{3}{2}$

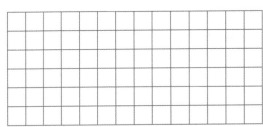

[3 marks]

C

FM 3 Alf the gas fitter uses this formula to work out how much to charge for a job:

$C = 20 + 30H$

where C is the charge and H is how long the job takes.

Bernice the gas fitter uses this formula:

$C = 30 + 25H$

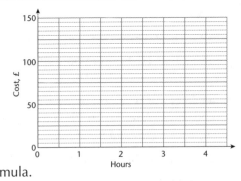

a On the grid draw lines to represent these formula. [2 marks]

b An engineer estimates that fitting a boiler will take between $1\frac{1}{2}$ and 4 hours.

Which fitter would be the best to employ to do the job?

Give a reason for your answer.

_____ [2 marks]

1 Here are the equations of six lines.

A $y = 3x + 6$ **B** $y = 2x - 1$ **C** $y = \frac{1}{2}x - 1$

D $y = 3x + 1$ **E** $y = \frac{1}{3}x + 1$ **F** $y = 4x + 2$

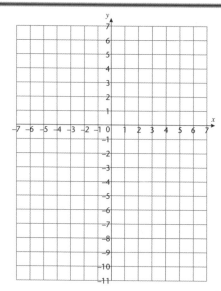

a Which line is parallel to line A?

_____ **[1 mark]**

b Which line crosses the y-axis at the same point as line B?

_____ **[1 mark]**

c Which other two lines intersect on the y-axis?

_____ and _____ **[1 mark]**

d Use the gradient-intercept method to draw the graph of $y = 3x - 2$ for $-3 \leqslant x \leqslant 3$ on the axes supplied.

[2 marks]

2 Use the cover-up method to draw these graphs on the axes supplied.

a $5x - 2y = 10$

b $2x - 3y = 6$

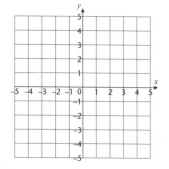

[2 marks]

AU 3 The diagram shows a hexagon ABCDEF.

The equation of the line through A and B is $x + y = 4$.

The equation of the line through D and C is $y = -3$.

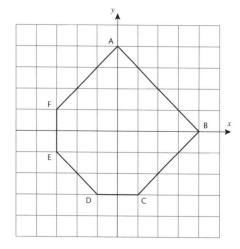

a Write down the equation of the lines through the points

　i A and F

_____ **[1 mark]**

　ii F and E.

_____ **[1 mark]**

b The gradient of the line through A and D is 7.

Write down the gradient of the lines through

　i A and C

_____ **[1 mark]**

　ii F and B.

_____ **[1 mark]**

C

1 a What is the equation of the line shown?

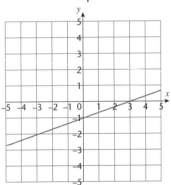

_____ [2 marks]

b On the same axes, draw the line $y = 2x + 4$.   [2 marks]

c Where do the graphs intersect?   _____ [1 mark]

B

FM 2 Frank compares two of his fuel bills.

For the first quarter of the year he used 200 units and paid £35.

For the second quarter of the year he used 150 units and paid £30.

a Plot this information on the axes below.

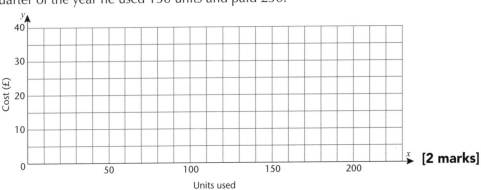

[2 marks]

b Join the points with a straight line and extend it to meet the axis.   [1 mark]

c Use the graph to establish a formula between the cost of fuel, C, and the number of units used, n.

_____ [2 marks]

B

FM 3 An online music store has different charges depending on how many tunes are downloaded.

This graph shows how much is charged for up to 20 downloads.

Work out the values of a, b, c, d, e and f to show these charges as equations:

$y = ax + b$ $0 < x \le 5$

$y = cx + d$ $5 < x \le 10$

$y = ex + f$ $12 < x \le 20$

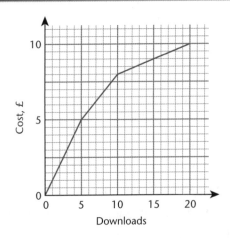

$a =$ _____ $b =$ _____ $c =$ _____ $d =$ _____ $e =$ _____ $f =$ _____ [3 marks]

Linear graphs and equations

1 **a** Draw the graph of $2x + 3y = 6$ on the axes opposite.

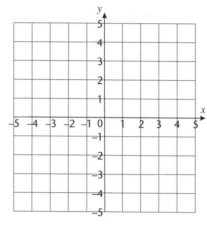

[2 marks]

b On the same axes, draw the line $y = 2x - 2$. **[2 marks]**

c Use the graph to find the solution to these simultaneous equations.

$2x + 3y = 6$

$y = 2x - 2$

$x =$ _____

$y =$ _____ **[1 mark]**

---

2 A is the point $(-4, -2)$ and B is the point $(5, 1)$.

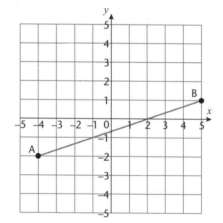

a What is the mid-point of AB?

_____ **[1 mark]**

b What is the gradient of AB?

_____ **[1 mark]**

c Find the equation of the line perpendicular to AB that passes through its mid-point.

_____ **[2 marks]**

AU 3 Here are the equations of three lines.

Line A: $y = 4x - 3$

Line B: $4y + x = 5$

Line C: $y = -\frac{1}{4}x - 3$

a Give a mathematical property shared by lines A and C. _____ **[1 mark]**

b Give a mathematical property shared by lines B and C. _____ **[1 mark]**

c Where do lines A and B intersect? _____ **[2 marks]**

Algebra

Quadratic graphs

C

1 **a** Complete the table of values for $y = x^2 - 2x + 1$.

x	−2	−1	0	1	2	3	4
y	9	4	1				

[1 mark]

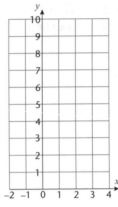

b Draw the graph of $y = x^2 - 2x + 1$ for values of x from −2 to 4.

[2 marks]

c Use the graph to find the value(s) of x when $y = 6$. _____ [1 mark]

d Use the graph to solve the equation $x^2 - 2x + 1 = 0$.

$x =$ _____ [1 mark]

C

2 **a** Complete the table of values for $y = x^2 + 2x - 1$.

x	−4	−3	−2	−1	0	1	2
y		2	−1				7

[2 marks]

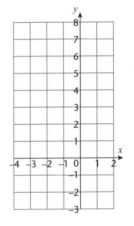

b Draw the graph of $y = x^2 + 2x - 1$ for values of x from −4 to 2.

[2 marks]

c Use the graph to find the value(s) of x when $y = 1.5$. _____ [1 mark]

d Use the graph to solve the equation $x^2 + 2x - 1 = 0$.

_____ [1 mark]

B

(AU 3) Here are the equations of three quadratic parabolas.

Parabola A: $y = 2x^2 - 3$

Parabola B: $y = x^2 + 1$

Parabola C: $y = x^2 - 3$

a Give a mathematical property that is shared by parabolas A and C.

_____ [1 mark]

b At what values of x do parabolas A and B intersect?

_____ [1 mark]

c Where do parabolas A and B intersect?

_____ [2 marks]

1 Below is the graph of $y = x^2 - x - 6$.

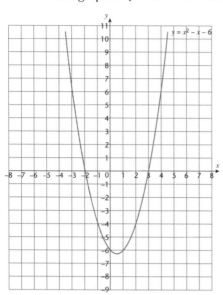

$y = x^2 - x - 6$

a Use the graph to solve
$x^2 - x - 6 = 0$.

$x = $ _____

$x = $ _____ **[1 mark]**

b Deduce the solutions to the equation
$x^2 - x - 12 = 0$.

$x = $ _____

$x = $ _____ **[1 mark]**

c By drawing a suitable straight line, solve
the equation $x^2 - 2x - 8 = 0$.

$x = $ _____

$x = $ _____ **[3 marks]**

2 Below is the graph of $y = 12x - x^3$ for values of x from -4 to 4.

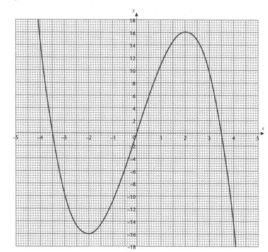

a Use the graph to solve $12x - x^3 = 0$. _____ **[1 mark]**

b By drawing a suitable straight line, solve the equation $11x - x^3 = 0$.

_____ **[3 marks]**

PS 3 Rhonda writes down the following facts about a quadratic equation.

- Roots are at $(-1, 0)$ and $(2, 0)$.

- Intercept on y-axis is $(0, -2)$.

- Vertex is at $(-\frac{1}{2}, -2\frac{1}{4})$.

a Explain, using a sketch, why Rhonda must be wrong.

_____ **[2 marks]**

b Three of the points that Rhonda states are correct.

Find the quadratic equation that Rhonda is solving. _____ **[2 marks]**

1   a   Complete the table of values for $y = (0.5)^x$.

x	0	1	2	3	4	5
y	1	0.5	0.25			0.03

[2 marks]

b   Draw the graph of $y = (0.5)^x$ for values of x from 0 to 5, on the axes opposite.

[2 marks]

c   Use your graph to solve the equation $0.5^x = 0.8$.

_____ **[1 mark]**

2   The sketch shows the graph of $y = x^3 + 2x^2 - 16x - 32$.

The expression $x^3 + 2x^2 - 16x - 32$ factorises to give $(x + 4)(x - 4)(x + 2)$.

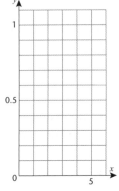

Write down the coordinates of:

A   _____ **[1 mark]**

B   _____ **[1 mark]**

C   _____ **[1 mark]**

D   _____ **[1 mark]**

PS 3   The graphs of $y = 2^x$ and $y = \dfrac{1}{x}$ are shown.

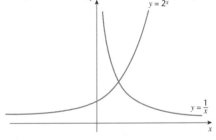

x	2^x	$y = \dfrac{1}{x}$
0.64	1.558	1.562
0.65	1.569	1.538

Complete the table to find, to 3 decimal places, the value of x where the two graphs intersect.

_____ **[3 marks]**

Algebraic fractions

1   a   Simplify $\dfrac{2x + 3}{2} + \dfrac{3x - 1}{3}$.

_____ **[2 marks]**

b   Simplify $\dfrac{3x - 1}{5} - \dfrac{2x - 5}{2}$.

_____ **[2 marks]**

c   Simplify $\dfrac{2x - 1}{3} \times \dfrac{4x - 1}{2}$.

_____ **[2 marks]**

d   Simplify $\dfrac{x - 1}{3} \div \dfrac{3x - 1}{6}$.

_____ **[2 marks]**

2   Simplify the following. Factorise and cancel where possible.

a $\dfrac{2x + 3}{5} \div \dfrac{6x + 9}{15}$

_____ **[2 marks]**

b $\dfrac{2x^2}{9} - \dfrac{2y^2}{3}$

_____ **[2 marks]**

PS 3   Fill in the missing expressions.

$\dfrac{3x}{\square} + \dfrac{\square}{2x} = \dfrac{6x^2 + y^3}{2xy}$

[2 marks]

1 a Solve the equation $\dfrac{x+1}{2} + \dfrac{x-3}{5} = 2$.

$x = $ _____ **[4 marks]**

b Solve the equation $\dfrac{4x+1}{3} - \dfrac{3x-1}{5} = 2$.

$x = $ _____ **[4 marks]**

c Solve the equation $\dfrac{3}{5x+1} + \dfrac{5}{3x-1} = 3$.

$x = $ _____ **[6 marks]**

2 a Show that the equation $\dfrac{3}{2x-1} - \dfrac{4}{3x-1} = 1$ simplifies to $x^2 - x = 0$.

_____ **[3 marks]**

b Hence, or otherwise, solve the equation $\dfrac{3}{2x-1} - \dfrac{4}{3x-1} = 1$.

$x = $ _____ **[2 marks]**

3 Solve this equation.

$$\dfrac{x}{x-1} + \dfrac{3}{x+1} = 1$$

$x = $ _____ **[5 marks]**

AU 4 An expression of the form $\dfrac{ax^2 + bx - c}{dx^2 - e}$ simplifies to $\dfrac{3x-1}{2x-1}$.

What was the original expression?

_____ **[3 marks]**

Simultaneous equations 2

1 Solve these simultaneous equations.

$y = 4 - x$

$x^2 + y = 16$

_____ **[5 marks]**

2 The diagram shows the line $2y = 3x - 13$ and the curve $x^2 + y^2 = 65$.

The curve and the line intersect at the points A and B.

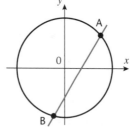

a Rearrange the equation $2y = 3x - 13$ to make y the subject.

_____ **[1 mark]**

b By solving the simultaneous equations $2y = 3x - 13$ and $x^2 + y^2 = 65$, find the coordinates of A and B.

_____ **[6 marks]**

3 Solve these simultaneous equations.

$x^2 + 2y^2 = 11$

$x = 3y$

_____ **[4 marks]**

AU 4 **a** Solve the simultaneous equations $y^2 + x^2 = 5$ and $y = 5 - 2x$.

_____ **[5 marks]**

b Which of the sketches below represents the graphs of the equations in part **a**?

Explain your choice.

I 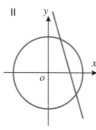   II 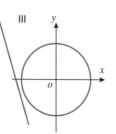   III

_____ **[1 mark]**

The nth term

C

AU 1 The nth term of a sequence is $4n + 1$.

    a Write down the first three terms of the sequence.

        _____ **[1 mark]**

    b Which term of the sequence is equal to 29?

        _____ **[1 mark]**

    c Explain why 84 is *not* a term in this sequence.

        _____

        _____ **[1 mark]**

C

2 What is the nth term of the sequence 3, 10, 17, 24, 31, ...?

        _____ **[2 marks]**

D

AU 3 R is an odd number. Q is an even number. P is a prime number.

    Tick the appropriate box to identify these expressions as *always even, always odd* or *could be either.*

	Always even	Always odd	Could be either	
a $R + Q$	☐	☐	☐	**[1 mark]**
b RQ	☐	☐	☐	**[1 mark]**
c $P + Q$	☐	☐	☐	**[1 mark]**
d R^2	☐	☐	☐	**[1 mark]**
e $R + PQ$	☐	☐	☐	**[1 mark]**

C

AU 4 **a** n is a positive integer. Explain why $2n$ is always an even number.

        _____

        _____ **[1 mark]**

    b Zoe says that when you square an even number you always get a multiple of 4.

        Show that Zoe is correct.

        _____

        _____ **[2 marks]**

B

PS 5 Sequence A has an nth term $2n + 3$.

    Sequence B has an nth term $5n - 1$.

    The first term that both sequences have in common is 9.

    Find the nth term of the sequence formed by all the common terms of the two sequences.

        _____ **[3 marks]**

1 Matches are used to make patterns with hexagons.

Pattern 1   **Pattern 2**   **Pattern 3**   **Pattern 4**

a Complete the table that shows the number of matches used to make each pattern.

Pattern number	1	2	3	4	5
Number of matches	6	11			

[1 mark]

b How many matches will be needed to make the 20th pattern?

_____ **[1 mark]**

c How many matches will be needed to make the nth pattern?

_____ **[2 marks]**

2 Rearrange the formula to make x the subject.

$$y = \frac{x + 2}{x - 4}$$

_____ **[4 marks]**

3 Rearrange the formula to make x the subject.

Simplify your answer as much as possible.

$$6x + 2y = 4(x + 3)$$

_____ **[3 marks]**

PS 4 $y = \dfrac{x + 1}{2x - 1}$ and $z = \dfrac{x + 3}{2x - 1}$.

Show that $z = \dfrac{7y - 2}{3}$.

_____ **[5 marks]**

AQA 2   EDEXCEL 3   OCR 2

1 a Solve the inequality $3x - 4 \leqslant 2$.

_____ **[2 marks]**

b What inequality is shown on this number line?

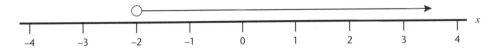

_____ **[1 mark]**

c Write down all the integers that satisfy both inequalities in parts **a** and **b**.

_____ **[1 mark]**

2 a What inequality is shown on this number line?

_____ **[1 mark]**

b Solve these inequalities.

i $\frac{x}{2} + 3 > 1$

_____ **[2 marks]**

ii $\frac{(x + 3)}{2} \leqslant 1$

_____ **[2 marks]**

c Write down all the integers that satisfy both inequalities in parts **b i** and **ii**.

_____ **[1 mark]**

3 Solve these inequalities.

a $3x - 2 \geqslant x + 7$

_____ **[3 marks]**

b $3(x - 1) < x - 5$

_____ **[3 marks]**

PS 4 What numbers are being described?

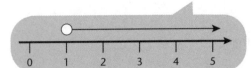

x is an odd number. $2x - 5 \leqslant 14$

_____ **[2 marks]**

B

1   On the graph, the region **R** is shaded.

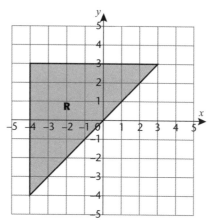

Write down the three inequalities that together describe the shaded region.

_____

_____

_____ **[3 marks]**

B

2   On the grid below, indicate the region defined by these three inequalities.

Mark the region clearly as **R**.

$x \leq 2$

$x + y > -3$

$y > -3$

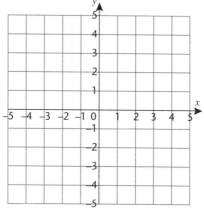

[3 marks]

AU 3 x and y are positive numbers. If $x + y < 30$, which of the following:

  • may be True (M)

  • must be false (F)

  • must be true (T)?

a $x = 40$ **b** $x + y \leq 20$ **c** $x - y = 10$

d $x > 35$ **e** $x + y = 30$ **f** $y < 30 - x$

g $y = 2x$ **h** $x + y \geq 25$

_____ **[3 marks]**

1 The graph of $y = \sin x$ is shown by a dotted line on the axes in parts **a** to **c** below.

Sketch the graphs stated in each case.

a $y = \sin x - 1$

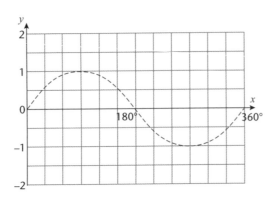

[1 mark]

b $y = \frac{1}{2} \sin x$

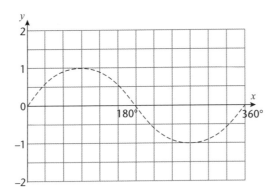

[1 mark]

c $y = \sin (x + 90°)$

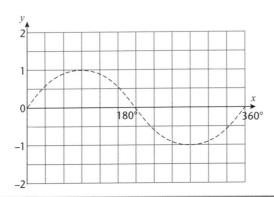

[1 mark]

AU 2 The graph $y = \cos x$ is transformed by $y = \cos (x + a)$.

After the transformation the graphs look exactly the same.

Write down a possible value of a. _____ [1 mark]

PS 3 The transformations $y = \sin (x + 90)$ and $y = \frac{1}{2} \sin x$ are applied to the graph of $y = \sin x$.

For the range $0 \leq x \leq 360$, how many solutions does the following equation have?

$\sin (x + 90) = \frac{1}{2} \sin x$

_____ [3 marks]

AQA 3   EDEXCEL 3   OCR 3

1 The graph of $y = \sin x$ is shown by a dotted line on the axes in parts **a** to **c** below.

Sketch the graphs stated in each case.

a $y = \sin 3x$

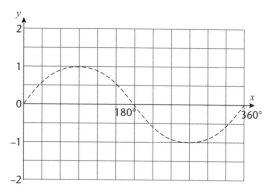

[1 mark]

b $y = -\sin x$

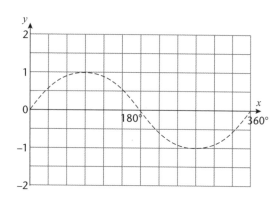

[1 mark]

c $y = -\sin x + 1$

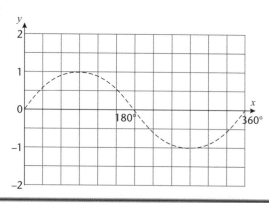

[1 mark]

AU 2 **a** Write the equation $y = x^2 + 2x - 6$ in the form $y = (x + a)^2 - b$.

_____ [2 marks]

b Describe the transformations that would need to be applied to the curve $y = x^2$ to map it to $y = x^2 + 2x - 6$.

_____ [2 marks]

Algebra

1 Prove that $(n + 5)^2 - (n + 3)^2 = 4(n + 4)$.

[3 marks]

2 ABCD is a cyclic quadrilateral.

Triangle ADB is an isosceles triangle where AB = AD.

Prove that AD is parallel to CB.

Give reasons for any angle values that you calculate or write down.

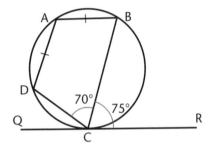

[3 marks]

3 The nth term of the triangular number sequence is given by $\frac{1}{2}n(n + 1)$.

Let any term in the triangular number sequence be T.

Prove that $8T + 1$ is always a square number.

[3 marks]

PS 4 OAB is a triangle such that $\overrightarrow{OA} = 2\mathbf{a}$ and $\overrightarrow{OB} = 2\mathbf{b}$.

M and N are the mid-points of $\overrightarrow{OA}$ and $\overrightarrow{OB}$ respectively.

X is the median of the triangle, which is where the lines from each vertex to the opposite mid-point meet.

Prove that the median, X, divides the lines AN and BM in the ratio 2 : 1.

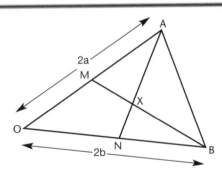

[4 marks]

Algebra record sheet

Name _____   Marks _____
Form _____   Percentage _____
Date _____   Grade _____

Page number	Question number	Topic	Mark	Comments
168	1	Basic algebra	/4	
	2	Basic algebra	/7	
	3	Basic algebra	/2	
	4	Basic algebra	/3	
169	1	Linear equations	/8	
	2	Linear equations	/9	
	3	Linear equations	/3	
170	1	Linear equations	/6	
	2	Linear equations	/3	
	3	Linear equations	/3	
171	1	Trial and improvement	/3	
	2	Trial and improvement	/4	
	3	Trial and improvement	/4	
172	1	Simultaneous equations	/10	
	2	Simultaneous equations	/4	
	3	Simultaneous equations	/4	
173	1	Simultaneous equations and formulae	/5	
	2	Simultaneous equations and formulae	/10	
	3	Simultaneous equations and formulae	/4	
174	1	Algebra 2	/6	
	2	Algebra 2	/4	
	3	Algebra 2	/7	
	4	Algebra 2	/2	
175	1	Factorising quadratic expressions	/4	
	2	Factorising quadratic expressions	/4	
	3	Factorising quadratic expressions	/4	
	4	Factorising quadratic expressions	/6	
	5	Factorising quadratic expressions	/3	
176	1	Solving quadratic equations	/3	
	2	Solving quadratic equations	/4	
	3	Solving quadratic equations	/6	
	4	Solving quadratic equations	/6	
	5	Solving quadratic equations	/4	
177	1	Completing the square	/4	
	2	Completing the square	/4	
	3	Completing the square	/3	
	4	Completing the square	/4	
	5	Completing the square	/2	
178	1	Solving problems with quadratic equations	/2	
	2	Solving problems with quadratic equations	/5	
	3	Solving problems with quadratic equations	/5	
	4	Solving problems with quadratic equations	/4	
179	1	Real-life graphs	/5	
	2	Real-life graphs	/3	
	3	Real-life graphs	/3	
180	1	Linear graphs	/2	
	2	Linear graphs	/6	
	3	Linear graphs	/4	
181	1	Linear graphs	/5	
	2	Linear graphs	/2	
	3	Linear graphs	/4	
182	1	Equations of lines	/5	

Page number	Question number	Topic	Mark	Comments
	2	Equations of lines	/5	
	3	Equations of lines	/3	
183	1	Linear graphs and equations	/5	
	2	Linear graphs and equations	/4	
	3	Linear graphs and equations	/4	
184	1	Quadratic graphs	/5	
	2	Quadratic graphs	/6	
	3	Quadratic graphs	/4	
185	1	Quadratic graphs	/5	
	2	Quadratic graphs	/4	
	3	Quadratic graphs	/4	
186	1	Other graphs	/5	
	2	Other graphs	/4	
	3	Other graphs	/3	
187	1	Algebraic fractions	/8	
	2	Algebraic fractions	/4	
	3	Algebraic fractions	/2	
188	1	Solving equations	/14	
	2	Solving equations	/5	
	3	Solving equations	/5	
	4	Solving equations	/3	
189	1	Simultaneous equations 2	/5	
	2	Simultaneous equations 2	/7	
	3	Simultaneous equations 2	/4	
	4	Simultaneous equations 2	/6	
190	1	The nth term	/3	
	2	The nth term	/2	
	3	The nth term	/5	
	4	The nth term	/3	
	5	The nth term	/3	
191	1	Formulae	/4	
	2	Formulae	/4	
	3	Formulae	/3	
	4	Formulae	/5	
192	1	Inequalities	/4	
	2	Inequalities	/6	
	3	Inequalities	/6	
	4	Inequalities	/2	
193	1	Graphical inequalities	/3	
	2	Graphical inequalities	/3	
	3	Graphical inequalities	/3	
194	1	Graph transforms	/3	
	2	Graph transforms	/1	
	3	Graph transforms	/3	
195	1	Graph transforms	/3	
	2	Graph transforms	/4	
196	1	Proof	/3	
	2	Proof	/3	
	3	Proof	/3	
	4	Proof	/4	
Total			**/441**	

Successes

1 _____

2 _____

3 _____

Areas for improvement

1 _____

2 _____

3 _____

Area of trapezium $= \frac{1}{2}(a + b)h$

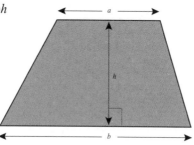

Volume of prism = area of cross-section × length

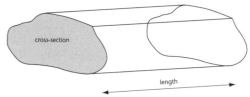

Volume of a sphere $= \frac{4}{3}\pi r^3$

Surface area of a sphere $= 4\pi r^2$

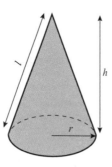

Volume of a cone $= \frac{1}{3}\pi r^2 h$

Curved surface area of a cone $= \pi r l$

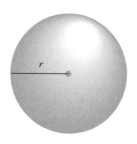

In any triangle: ABC

Area of triangle $= \frac{1}{2}ab \sin C$

Sine rule $\dfrac{a}{\sin A} = \dfrac{b}{\sin B} = \dfrac{c}{\sin C}$

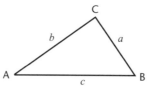

Cosine rule $a^2 = b^2 + c^2 - 2bc \cos A$

The quadratic equation

The solutions for $ax^2 + bx + c = 0$, where $a \neq 0$, are given by $x = \dfrac{-b \pm \sqrt{(b^2 - 4ac)}}{2a}$

Glossary

alternate segment The other segment. The segment on the other side of the chord.

angle bisector A straight line or plane that divides an angle in half.

angle of rotation The angle turned through to move from one direction to another.

anticlockwise Turning in the opposite direction to the movement of the hands of a clock. (Opposite of clockwise.)

approximate An inexact value that is accurate enough for the current situation.

arc A curve forming part of the circumference of a circle.

area Measurement of the flat space a shape occupies. Usually measured in square units or hectares. (*See also* surface area *and* curved surface area.)

area ratio The ratio of the areas of two similar shapes is equal to the ratio of the squares of their corresponding lengths. The area ratio, or area scale factor, is the square of the length ratio.

area scale factor (*See* area ratio.)

area sine rule The area of a triangle is given by $\frac{1}{2}ab$SinC.

asymptote A straight line whose perpendicular distance from a curve tends to zero as the distance from the origin increases without limit.

average A single number that represents or typifies a collection of values. The three commonly used averages are mode, mean, and median.

balance the coefficients The first step to solving simultaneous equations is to make the coefficients of one of the variables the same. This is called balancing the coefficients.

best buy A purchase that gives best value for money spent.

boundary When drawing a graph of an inequality, first consider the equality. This line forms the limit or boundary of the inequality. When a strict inequality is stated (< or >), the boundary line should be drawn as a dashed line to show that it is not included in the range of values. When ≤ or ≥ are used, the boundary line should be drawn as a solid line to show that the boundary is included.

boundary line A line that defines an inequality when it is depicted in graph form.

box plot Or box-and-whisker plot. This is a way of displaying data for comparison. It requires five pieces of data; lowest value, the lower quartile (Q1), the median (Q2), the upper quartile (Q3) and the highest value.

brackets The symbols '(' and ')' which are used to separate part of an expression. This may be for clarity or to indicate a part to be worked out individually. When a number or value is placed immediately before an expression or value inside a pair of brackets, the two are to be multiplied together. For example, $6a(5b + c) = 30ab + 6ac$.

cancel A fraction can be simplified to an equivalent fraction by dividing the numerator and denominator by a common factor. This is called cancelling.

centre of enlargement The fixed point of an enlargement. The distance of each image point from the centre of enlargement is the distance of object point from centre of enlargement × scale factor. (*See also* scale factor.)

centre of rotation The fixed point around which a shape is rotated or turned.

check Calculations can be checked by carrying out the inverse operation. Solutions to equations can be checked by substituting values of the variable(s).

chord A line joining two points on the circumference of a circle.

circle A circle is the path of a point that is always equidistant from another point (the centre).

circle theorem A theorem proving a property of circles.

circumference The outline of a circle. The distance all the way around this outline.

clockwise Turning in the same direction as the movement of the hands of a clock. (Opposite of anticlockwise.)

coefficient The number in front of an unknown quantity (the letter) in an algebraic term. For example, in $8x$, 8 is the coefficient of x.

combined events Two or more events (independent or mutually exclusive) that may occur during a trial.

complementary event The complement of 'Event A happening' is 'Event A not happening'.

completing the square A method of solving quadratic equations which involves rewriting the equation $x^2 + px + q$ in the form $(x + a)^2 + b$.

compound interest Instead of interest being calculated once at the end of the term, it is calculated annually (or monthly or daily) and added to the Principal before the next period starts. This has the effect of increasing the amount of interest earned or owed above that calculated by using simple interest.

conditional probability This describes the situation when the probability of an event is dependent upon the outcome of another event. For example, the probability of the colour of a second ball drawn from a bag is conditional to the colour of the first ball drawn from the bag, if the first ball is not replaced.

cone A pyramid with a circular base.

congruent triangles Triangles that are exactly alike in shape and size.

construct To draw angles, lines or shapes accurately, according to given requirements.

continuous data Data that can be measured rather than counted, such as weight and height. It can have an infinite number of values within a range.

correlation The extent of correspondence between two variables. If the relation between two variables follows a consistent trend, then they are said to correlate.

cosine The ratio of the adjacent side to the hypotenuse in a right-angled triangle.

cosine rule A formula that enables you to find the lengths of sides or the size of an angle in a triangle: $a^2 = b^2 + c^2 - 2bc$CosA.

cover-up method To plot a straight line graph, only two points need to be known. (Finding a third point acts as a check.) The easiest points to find are those when $x = 0$ and $y = 0$. By covering over the x and y terms, you can easily *see* what the associated values of y and x are.

cross-section The shape of a slice through a solid.

cubic A cubic expression or equation contains an 'x^3' term.

cube The result of raising a number by the power of three. For example, 'two cubed' is written: 2^3, which is $2 \times 2 \times 2 = 8$.

cube root The cube root of a number is the value that must be multiplied by itself three times to get the number. For example, the cube root of 27 is 3 (written $\sqrt[3]{27} = 3$), because $3 \times 3 \times 3 = 3^3 = 27$. (*See also* square root.)

cubic graph The graph of an equation that includes an x^3 term.

cumulative frequency diagram A graph where the value of the variable is plotted against the total frequency **so far**. The last plot will show the total frequency.

curved surface area The area of the surface of a curved shape, such as a cylinder or sphere. For example, the area of paper needed to fold to make a cone or the label for a tin of baked beans.

cylinder A solid or hollow prism with circular ends and uniform (unchanging) cross-section. The shape of a can of baked beans or length of drainpipe.

data collection sheet A form or table which is used for recording data collected during a survey.

data handling cycle The process of conducting a survey. It consists of four parts: outlining the problem, stating how the data will be collected, stating how the data will be processed, and drawing a conclusion.

decimal place Every digit in a number has a place value (hundreds, tens, ones, etc.). The places after (to the right of) the decimal point have place values of tenths, hundredths, etc. These are called the decimal places.

denominator The number under the line in a fraction. It tells you the denomination, name or family of the fraction. For example, a denominator of 3, tells you you are thinking about thirds; the whole thing has been divided into three parts. (*See also* numerator.)

density The ratio of the mass of an object to its volume. The mass per unit volume.

diameter A straight line across a circle, from circumference to circumference and passing through the centre. It is the longest chord of a circle and two radii long. (*See also* radius.)

difference of two squares The result of $a^2 - b^2 = (a - b)(a + b)$ is called the difference of two squares.

direct proportion Two values or measurements may vary in direct proportion. That is, if one increases, then so does the other.

direct variation Another name for direct proportion.

direction The way something is facing or pointing. Direction can be described using the compass points (north, south, south-east, etc.) or using bearings (the clockwise angle turned from facing north). The direction of a vector is given by the angle it makes with a line of reference. Two vectors that differ only in magnitude will be parallel.

discrete data Data that is counted, rather than measured, such as favourite colour or a measurement that occurs in integer values, such as a number of days. It can only have specific values within a range.

discriminant The quantity $(b^2 - 4ac)$ in the quadratic formula is called the discriminant.

dispersion A measure of the spread of a set of data, for example, interquartile range.

distance The separation (usually along a straight line) of two points.

distance-time graph A graph showing the variation of the distance of an object from a given point during an interval of time. Usually, time is measured along the horizontal axis, and distance is measured up the vertical axis.

do the same to both sides To keep an equation balanced, you must do the same thing to both sides. If you add something to one side, you must add the same thing to the other side. If you double one side, you must double the other side, etc. If you are manipulating a fraction, you must do the same thing to the numerator and the denominator to keep the value of the fraction unchanged. However you can only multiply or divide the numbers. Adding or subtracting will alter the value of the fraction.

eliminate To get rid of.

enlargement A transformation of a plane figure or solid object that changes the size of the figure or object by a scale factor but leaves it the same shape.

equation A number sentence where one side is equal to the other. An equation always contains an equals sign (=).

equivalent fractions Fractions that by a process of cancellation, can be shown to represent the same rational number. For example, $\frac{3}{4}, \frac{6}{8}, \frac{9}{12}$, and $\frac{1}{28}$ are all equivalent fractions.

exact values An alternative name for surds.

exhaustive events The events giving all possible outcomes of a trial, so that their probabilities add up to 1.

expand Make bigger. Expanding brackets means you must multiply the terms inside a bracket by the number or letters outside. This will take more room to write, so you have 'expanded' the expression.

expand and simplify An expression involving brackets can be expanded (*see* expand) and the resulting terms may contain like terms that can then be simplified.

expectation Something you expect to happen.

exponent A power or index. For example, in the expression 3^4, 4 is the exponent, power or index.

exponential functions Equations which have the form $y = k^x$, where k is a positive number, are called exponential functions.

expression Collection of symbols representing a number. These can include numbers, variables (x, y, etc.), operations ($+$, $\times$, etc.), functions (squaring, cosine, etc.), but there will be no equals sign (=).

exterior angle Angles of a polygon that lie outside the shape when a side is produced. An exterior angle and its adjacent interior angle add up to $180°$.

factor A whole number that divides exactly into a given number.

factorisation Finding one or more factors of a given number or expression.

factorise (*See* factorisation.)

formula (plural: formulae) An equation that enables you to convert or find a measurement from another known measurement or measurements. For example, the conversion formula from the Fahrenheit scale of temperature to the more common Celsius scale is $\frac{C}{5} = \frac{(F - 32)}{9}$ where C is the temperature on the Celsius scale and F is the temperature on the Fahrenheit scale.

fraction A fraction means 'part of something'. To give a fraction a name, such as 'fifths' we divide the whole amount into **equal** parts (in this case five equal parts). A 'proper' fraction represents an amount less than one (the numerator is smaller than the denominator). Any two numbers or expressions can be written as a fraction, i.e. they are written as a numerator and denominator. (*See also* numerator *and* denominator.)

frequency density Frequency density = frequency of class interval ÷ width of class interval. This is represented on the vertical axis of a histogram.

frequency polygon A line graph drawn from the information given in a frequency table.

frequency table A table showing values (or classes of values) of a variable alongside the number of times each one has occurred.

function A function of x is any algebraic expression in which x is the only variable. This is often represented by the function notation f(x) or 'function of x'.

gradient How steep a hill or the line of a graph is. The steeper the slope, the larger the value of the gradient. A horizontal line has a gradient of zero.

gradient-intercept The point at which the gradient of a curve or line crosses an axis.

guess Using your mathematical knowledge, you can make an estimate of an answer and use this as a starting point in a trial and improvement problem.

highest common factor (HCF) When all the factors of two or more numbers are found, some numbers will have factors in common (the same factors). For example, 6 has factors 1, 2, 3 and 6. 9 has factors 1, 3, and 9. They both have the factors 1 and 3. (1 and 3 are common factors.) The greatest of these is 3, so 3 is the highest common factor.

histogram A diagram, similar to a bar chart, but where quantities are represented by rectangles of different, appropriate areas.

hypotenuse The longest side of a right-angled triangle. The side opposite the right angle.

hypothesis A theory or idea.

identity An identity is similar to an equation, but is true for all values of the variable(s). Instead of the usual equals (=) sign, $\equiv$ is used. For example, $2x \equiv 7x - 5x$.

image In geometry the 'image' is the result of a transformation.

independent events Two events are independent if the occurrence of one has no influence on the occurrence of the other. For example, getting a head when tossing a coin has no influence over scoring a 2 with a die. Missing a bus and getting to school on time are not independent events.

inequality An inequality shows two numbers or expressions that are not equal. Instead of the equals (=) sign, the symbol $\neq$ is used. If it is known which number is the greater or smaller, $>$, $<$, $\geqslant$ or $\leqslant$ could be used.

intersection A point or points common to two or more geometrical figures.

intercept The point where a line or graph crosses an axis.

interior angle of regular polygon An internal angle of a polygon. The interior angles of a regular polygon are equal.

interquartile range The difference between the values of the lower and upper quartiles.

inverse variation Another name for inverse proportion.

isosceles triangle A triangle with two sides that are equal. It also has two equal angles.

kite A quadrilateral with two pairs of adjacent sides that are equal. The diagonals of a kite are perpendicular, but only one of them bisects the kite.

length ratio The ratio of lengths in similar figures.

limit of accuracy No measurement is entirely accurate. The accuracy depends on the tool used to measure it. The value of every measurement will be rounded to within certain limits. For example, you can probably measure with a ruler to the nearest half-centimetre. Any measurement you take could be inaccurate by up to half a centimetre. This is your limit of accuracy.

line of best fit When data from an experiment or survey is plotted on graph paper, the points may not lie in an exact straight line or smooth curve. You can draw a line of best fit by looking at all the points and deciding where the line should go. Ideally, there should be as many points above the line as there are below it.

linear equation An equation in the first degree of the form $y = ax + b$, where x is limited to the power of 1. The graph of a linear equation is a straight line.

linear graphs A straight-line graph from an equation such as $y = 3x + 4$.

linear scale factor Also called the scale factor or length ratio. The ratio of corresponding lengths in two similar shapes is constant.

loci (See locus.)

locus (plural: loci) The locus of a point is the path taken by the point following a rule or rules. For example, the locus of a point that is always the same distance from another point is the shape of a circle.

long division A division involving numbers with a large number of digits.

long multiplication A multiplication involving numbers with a large number of digits.

lowest common multiple (LCM) Every number has an infinite number of multiples. Two (or more) numbers might have some multiples in common (the same multiples). The smallest of these is called the lowest common multiple. For example, 6 has multiples of 6, 12, 18, 24, 30, 36, etc. 9 has multiples of 9, 18, 27, 36, 45, etc. They both have multiples of 18 and 36, 18 is the LCM.

magnitude Size. Magnitude is always a positive value.

mass The mass is the amount of 'stuff' an object consists of. It does not vary if the object is moved somewhere else. The weight of an object is closely related to mass but depends on the effect of gravity. The weight of an object will be different on the Earth to what it is on the moon. Its mass will remain constant.

mean The mean value of a sample of values is the sum of all the values divided by the number of values in the sample. The mean is often called the average, although there are three different concepts associated with 'average': mean, mode and median.

median The middle value of a sample of data that is arranged in order. For example, the sample 3, 2, 6, 2, 3, 7, 4 may be arranged in order as follows 2, 2, 3, 3, 4, 6, 7. The median is the fourth value, which is 3. If there is an even number of values, the median is the mean of the two middle values, for example, 2, 3, 6, 7, 8, 9, has a median of 6.5.

mirror line A line where a shape is reflected exactly on the other side.

mixed number A number written as a whole number and a fraction. For example, the improper fraction $\frac{5}{2}$ can be written as the mixed number $2\frac{1}{2}$.

modal class Where data is grouped into classes, it is the class with the highest frequency.

mode The value that occurs most often in a sample. For example, the mode of the sample 2, 2, 3, 3, 3, 3, 4, 5, 5, is 3.

multiplier The number used to multiply by.

multiplier method A method of finding the original amount when given the value after a percentage change in the original amount (see reverse percentage).

mutually exclusive If the occurrence of a certain event means that another event cannot occur, the two events are mutually exclusive. For example, if you miss a bus you cannot also catch the bus.

negative (in maths) Something less than zero. The opposite of positive. (See also negative gradient and positive.)

negative correlation If the effect of increasing one measurement is to decrease another, they are said to show negative correlation. For example, the time taken for a certain journey will have negative correlation with the speed of the vehicle. The slope of the line of best fit has a negative gradient.

negative gradient When a line slopes down from left to right it has a negative gradient.

negative indices Entities with negative indices denote the reciprocal of the entities with positive indices. For example x^{-4} is the same as $\frac{1}{x^4}$.

negative reciprocal The reciprocal multiplied by –1. The gradients of perpendicular lines are the negative reciprocal of each other.

non-linear An expression or equation that does not form a straight line. The highest power of x, is greater than 1.

*n***th term** The 'general' term in a sequence. The formula for the nth term describes the rule used to get any term.

number line A line where all the numbers (whole numbers and fractions) can be shown by points at distances from zero.

numerator The number above the line in a fraction. It tells you the number of parts you have. For example, $\frac{3}{5}$ means you have three of the five parts. (*See also* denominator.)

object (in maths) You carry out a transformation on an object to form an image. The object is the original or starting shape, line or point.

ordered data Data or results arranged in ascending or descending order.

parabola The shape of a graph plotted from an equation such as $5x^2 - 7x + 2 = 0$. The equation will have an 'x^2' term.

parallel Straight lines that are always the same distance apart.

parallelogram A four-sided polygon with two pairs of equal and parallel opposite sides.

pattern A repetition or connection occurring between shapes or numbers.

percentage decrease If an actual amount decreases, the percentage change will be a percentage decrease.

percentage increase If an actual amount increases, the percentage change will be a percentage increase.

percentage multiplier Percentage expressed as a decimal. For example, 23% would be represented by 0.23. An increase of 8% is a multiplier of 1.08. A decrease of 11% is a multiplier of 0.89.

perimeter The outside edge of a shape. The distance around the edge. The perimeter of a circle is called the circumference.

perpendicular At right angles (90°) to a given line or surface.

perpendicular bisector A line drawn at a right angle to a line segment which also divides it into two equal parts.

point of contact The point where a tangent touches a circle.

polygon A closed shape with three or more straight sides.

population All the members of a particular group. The population could be people or specific outcomes of an event.

positive (in maths) Something greater than zero. The opposite of negative. (*See also* positive gradient *and* negative.)

positive correlation If the effect of increasing one measurement is to increase another, they are said to show positive correlation. For example, the time taken for a journey in a certain vehicle will have positive correlation with the distance covered.

positive gradient When a line slopes up from left to right it has a positive gradient.

power The number of times a number or expression is multiplied by itself. The name given to the symbol to indicate this, such as 2. (*See also* square *and* cube.)

powers of 10 All the numbers 10^n; 1, 10, 100...

primary data Data in its original form as collected in experiments or surveys.

prime factor A factor of a number that is also a prime number. (*See also* prime number.)

prime number A number whose only factors are 1 and itself. 1 is not a prime number. 2 is the only even prime number.

prism A 3-D shape whose ends are identical shapes and the other sides are perpendicular to the ends. A box is a rectangular prism. The volume of a prism is calculated by multiplying the area of the uniform cross-section by the length.

proof An argument that establishes a fact about numbers or geometry for all cases. Showing the fact is true for specific cases is a demonstration.

pyramid A polyhedron on a triangular (*see* triangle), square, or polygonal (*see* polygon) base, with triangular faces meeting at a vertex. The volume, V, of a pyramid of base area A and perpendicular height h, is given by the formula $V = \frac{1}{3}A \times h$.

Pythagoras' theorem The theorem states that the square of the hypotenuse of a right-angled triangle is equal to the sum of the squares of the other two sides.

quadratic An expression, equation or formula involving an 'x^2' term.

quadratic equation An equation that includes an x^2 term. It has the form $y = ax^2 + bx + c$.

quadratic expansion Expanding two brackets $(x + a)(x + b)$ to give a quadratic expression.

quadratic expression An expression involving an x^2 term.

quadratic formula A formula for solving quadratic equations:
$$x = \frac{-b \pm \sqrt{b^2 - 4ac}}{2a}$$

qualitative data Data that is imprecise or approximate, or data that is not expressed numerically such as the colour of cars or flavour of food. It represents the converse of quantitative data.

quantitative data Data that is determined by measurement and can be expressed numerically.

questionnaire A list of questions distributed to people so statistical information can be collected.

radius (plural: radii) The distance from the centre of a circle to its circumference.

random A random number is one chosen without following a rule. Choosing items from a bag without looking means they are chosen at random; every item has an equal chance of being chosen. People selected for a survey may be chosen at random.

range The difference between the smallest and largest values in a set of data. It is a measure of the dispersion of the data.

ratio The ratio of A to B is a number found by dividing A by B. It is written as A : B. For example, the ratio of 1 m to 1 cm is written as 1 m : 1 cm = 100 : 1. Notice that the two quantities must both be in the same units if they are to be compared in this way.

rational number A rational number is a number that can be written as a fraction, for example, $\frac{1}{4}$ or $\frac{10}{3}$.

rationalise Removing a surd from a denominator (by multiplying the numerator and denominator by that surd) is called rationalising the denominator.

raw data Data in the form it was collected. It hasn't been ordered or arranged in any way.

reciprocal The reciprocal of any number is 1 divided by the number. The effect of finding the reciprocal of a fraction is to turn it upside down. The reciprocal of 3 is $\frac{1}{3}$, the reciprocal of $\frac{1}{4}$ is 4, the reciprocal of $\frac{10}{3}$ is $\frac{3}{10}$.

rectangle A quadrilateral in which all the interior angles are 90°. The opposite sides are of equal length. When all four sides are of equal length it is called a square.

reflection The image formed after being reflected. The process of reflecting an object.

region An area on a graph defined by certain rules or parameters.

regular polygon A polygon that has sides of equal length and angles of equal size.

relative frequency Also known as experimental probability. It is the ratio of the number of successful events to the number of trials. It is an estimate for the theoretical probability.

reverse percentage The process of finding the original amount when given the value after a percentage change in the original amount.

rhombus A parallelogram that has sides of equal length. A rhombus has two lines of symmetry and a rotational symmetry of order 2. The diagonals of a rhombus bisect each other at right angles and they bisect the figure.

roots The roots of a quadratic equation are the values of x when $y = 0$. (They are the solution to the equation.) They can be seen on a graph where the parabola crosses the x-axis.

rotation Turn. A geometrical transformation in which every point on a figure is rotated through the same angle about the same centre.

round To approximate a number so that it is accurate enough for some specific purpose. The rounded number may be used to make arithmetic easier and is always less precise than the unrounded number.

rule An alternative name for a formula.

sample The part of a population that is considered for statistical analysis. The act of taking a sample within a population is called sampling. There are two factors that need to be considered when sampling from a population: 1. The size of the sample. The sample must be large enough for the results of a statistical analysis to have any significance. 2. The way in which the sampling is done. The sample should be representative of the population.

scale factor The ratio by which a length or other measurement is increased or decreased.

scale A scale on a diagram shows the scale factor used to make the drawing. The axes on a graph or chart will use a scale depending on the space available to display the data. For example, each division on the axis may represent 1, 2, 5, 10 or 100, etc. units.

scatter diagram A diagram of points plotted of pairs of values of two types of data. The points may fall randomly or they may show some kind of correlation.

secondary data Data that is collected from existing sources such as lists or tables.

sector A region of a circle, like a slice of a pie, bounded by an arc and two radii.

segment A part of a circle between a chord and the circumference.

significant figures The significance of a particular digit in a number is concerned with its relative size in the number. The first (or most) significant figure is the left-most, non-zero digit; its size and place value tell you the approximate value of the complete number. The least significant figure is the right-most digit; it tells you a small detail about the complete number. For example, if we write 78.09 to 3 significant figures we would use the rules of rounding and write 78.1.

similar The same shape but a different size.

similar triangles Triangles with the same size angles. The lengths of the sides of the triangles are different but vary in a constant proportion.

simultaneous equations Two or more equations that are true at the same time.

sine The ratio of the opposite side to the hypotenuse in a right-angled triangle.

sine rule In a triangle, the ratio of the length of a side to the sine of the opposite angle is constant, hence $\frac{a}{\sin A} = \frac{b}{\sin B} = \frac{c}{\sin C}$.

slant height The distance along the sloping edge of a cone or pyramid.

solve Finding the value or values of a variable (x) which satisfy the given equation or problem.

special sequences Sequences that occur frequently. For example, even numbers, odd numbers, triangular numbers and prime numbers.

speed The rate at which something moves or happens.

sphere A sphere is a ball shape. For example, the moon and the planets are spheres.

square 1. A polygon with four equal sides and all the interior angles equal to 90°.

square 2. The result of multiplying a number by itself. For example, 5^2 or 5 squared is equal to $5 \times 5 = 25$.

square root The square root of a number is the value that must be multiplied by itself to get the number. For example, the square root of 9 is 3 (written $\sqrt{9} = 3$), because $3 \times 3 = 3^2 = 9$. (*See also* cube root.)

standard form Also called standard index form. Standard form is a way of writing very large and very small numbers using powers of 10. A number is written as a value between 1 and 10 multiplied by a power of 10. That is, $a \times 10^n$ where $1 \leqslant a < 10$, and n is a whole number.

stem-and-leaf diagram An ordered representation of classes of statistical data. The classes are visualised as the stem of a plant and the data points as its leaves.

subject The subject of a formula is the letter on its own on one side of the equals sign. For example, t is the subject of this formula: $t = 3f + 7$.

substitute When a letter in an equation, expression or formula is replaced by a number, we have substituted the number for the letter. For example, if $a = b + 2x$, and we know $b = 9$ and $x = 6$, we can write $a = 9 + 2 \times 6$. So $a = 9 + 12 = 21$.

surd A number written as $\sqrt{x}$. For example $\sqrt{7}$.

surd form The square root sign is left in the final expression when $\sqrt{x}$ is an irrational number.

surface area The area of the surface of a 3-D shape, such as a prism. The area of a net will be the same as the surface area of the shape.

survey A questionnaire or interview held to find data for statistical analysis.

tangent A straight line that touches the circumference of a circle at one point only.

tetrahedron A solid (3-D) figure with four triangular faces. If the faces are equilateral triangles it is called a *regular* tetrahedron.

time How long something takes. Time is measured in days, hours, seconds, etc.

top-heavy A fraction where the numerator is greater than the denominator.

transform Change.

transformation An action such as translation, reflection or rotation.

translation A transformation in which all points of a plane figure are moved by the same amount and in the same direction.

trapezium A quadrilateral with one pair of parallel sides.

tree diagram A diagram to show all the possible outcomes of combined events.

trial and improvement Some problems require a knowledge of mathematics beyond GCSE level but sometimes these can be solved or estimated by making an educated guess and then refining this to a more accurate answer. This is known as trial and improvement.

trigonometry The branch of mathematics concerned with trigonometric functions (sine, cosine, tangent, etc.) and their application to triangles.

two-way tables These link two variables. One is listed in the column headers and the other in the row headers; the combination of the variables is shown in the body of the table.

unitary method A method of calculation where the value for one item is found before finding the value for several items.

unordered data (*See* raw data.)

variable A quantity that can have many values. These values may be discrete or continuous. They are often represented by x and y in an expression.

vector A quantity with magnitude **and** direction.

vertex 1. The points at which the sides of a polygon or the edges of a polyhedron meet.

vertex 2. The turning point (maximum or minimum) of a graph.

volume The amount of space occupied by a substance or object or enclosed within a container.

volume ratio The ratio of the volumes of two similar shapes is equal to the ratio of the cubes of their corresponding lengths. The volume ratio, or volume scale factor, is the cube of the length ratio.

volume scale factor (*See* volume ratio.)

x-value The value along the horizontal axis on a graph using the Cartesian coordinate system.

$y = mx + c$ The general equation of a straight line. m is the gradient and c is the y-intercept.

y-value The value along the vertical axis on a graph using the Cartesian coordinate system.

π (pronounced 'pi') The numerical value of the ratio of the circumference of a circle to its diameter (approximately 3.14159).

Answers

In this mark scheme there are five types of mark.

B marks: These are for questions which are worth only 1 mark, or where there is no method needed.

M marks: These are for correct methods that would lead to the answer.

A marks: These are accuracy marks for correct answers.

E marks: These are for explanations.

Q marks: These are for the quality of written communication.

You will also see other abbreviations:

ft   This means 'follow through' meaning that your answer can be marked correct if you have used an earlier incorrect answer but done all the subsequent working correctly.

112   Statistics 1

Q	Answer	Mark	Comment
1a	2	B1	Remember, the mode is the number with the biggest frequency.
1b	2	B1	Add up the frequencies until a total of 50 is passed.
1c	$0 \times 8 + 1 \times 23 + 2 \times 52 + 3 \times 15 + 4 \times 2$ (= 180)	M1	
	$180 \div 100$	M1	
	1.8	A1	
2a	0	B1	It is always worth doing a total of the rows and columns.
2b	1	B1	
2c	$0 \times 11 + 1 \times 15 + 2 \times 18 + 3 \times 5 + 4 \times 1$ (= 70)	M1	
	$70 \div 50$	M1	
	1.4	A1	
2d	The top half of the table has bigger numbers or mean number of sisters 1.14 < mean number of brothers	E1	
3	5, 8, 9, 9, 9	B3	9 must occur at least twice to be the mode. The total of all five must be 40. Two 9s leaves 22 which can only be made by 7, 7 and 8. So try three 9s which leaves 13. Only 5, 8 gives the necessary range. B2 Two of the conditions are met, e.g. 6, 7, 9, 9, 9. B1 One of the conditions is met, e.g. total of five numbers is 40.

113   Statistics 1

Q	Answer	Mark	Comment
1a	$27 \times 45 + 39 \times 55 + 78 \times 65 + 31 \times 75 + 13 \times 85 + 12 \times 95$ (= 13 000)	M1	
	$13\ 000 \div 200$	M1	
	65	A1	
1b		B1	All points must be plotted correctly.
1c	Girls did better; polygons are about the same shape and girls are 10 marks better	E1	Do not just say 'girls did better', always refer to a numerical value.
2a	Any frequency ÷ class width	M1	To find the frequency density, divide each frequency by the class width.
	1.5, 4.8, 7.2, 1.8, 0.35	A1	

2b		B2ft	If you made an error calculating the frequency densities but plotted your values correctly you will still get full marks. You will lose 1 mark for every error.
3	Attempt to find areas of first three bars	M1	Pick any scale for the side axis, such as 1, 2 etc, and work out the area.
	Frequency density scale marked in 2's.	A1	Compare your area to 115 to get the actual scale.
	Working out frequency of each bar: 30, 40, 45, 55, 70, 45, 60	M1	Work out the area of each bar.
	75	A1	Count the area above 55

114 Statistics 1

Q	Answer	Mark	Comment
1a	<table><tr><td></td><td>Boys</td><td>Girls</td></tr><tr><td>0 < time (hours) ≤ 4</td><td></td><td></td></tr><tr><td>4 < time (hours) ≤ 6</td><td></td><td></td></tr><tr><td>6 < time (hours) ≤ 8</td><td></td><td></td></tr><tr><td>8 < time (hours) ≤ 10</td><td></td><td></td></tr><tr><td>More than 10 hours</td><td></td><td></td></tr></table>	B2	You will get 1 mark for listing boys/girls and 1 mark for the times, but make sure that the times do not overlap and that no times are missed out.
1b	Not really; the difference is not that large and the sample of girls was too small	B1	
2	$39 + 45 + 32 + \dots (= 432)$	M1	The clue in the question is that the mean and range for 2008 is given. This suggests you should work out the mean and range for 2009.
	$432 \div 12$	M1	
	36	A1	
	Conclusion using data such as 'There is not enough evidence that the rainfall is increasing as the mean has only increased by 1 mm. The range for the two years is also much the same, at 38 and 40.'	E1 Q1	These marks are given for the overall conclusion. The E mark will be for stating values. The Q mark is for a clear conclusion.
3	A: Look at past historical data B: Test by experiment C: Find out how many tickets were sold (equally likely outcomes) D: Do a survey	B3	There is no need to write lots of words. A brief answer such as 'Experiment' will often do. You lose a mark for every one you get wrong.

115 Statistics 1

Q	Answer	Mark	Comment
1a	Leading question Double negative Two questions in one Only two responses	B2	You only need to give two of these answers.
1b	A set of at least three responses from £0 to a reasonable amount, say £2, with no overlaps and no gaps. e.g. $£0 < m \le 50p$, $50p < m \le £1$, $£1 < m \le £1.50$, $£1.50 < m \le £2$	B2	It is important to use common sense and your own experience in this type of question. You could say 'More than £2' for the last choice. You will lose 1 mark for each condition that is not met.
2a	Any two from: sample too small; sample not representative; sample not random; conclusion incorrect, rounds to 6 out of 10	B2	1 mark for each answer
2b	32 out of 50 is nearer 6 out of 10 and it is a small, unrepresentative sample	B1	
3	Showing a calculation such as $50 \div 366 \times 8$	M1	Show that you know how to calculate the number to be surveyed if there are 50 out of 366.
	Showing that the figures in the table are correct rounded values for the calculation	A1	The answers are not accurate so round off.
	One comment from: 50 is over 10% of company so a reasonable number to survey. 51 in survey not 50. Must have 1 female manager.	E1	To get exactly 50 one value will have to be reduced, but you must have at least one from each group and survey at least 10% to get a representative sample.

116 Statistics 2

Q	Answer	Mark	Comment
1a	10am–11am	B1	
1b	11am	B1	
1c	25 °C	B1	

1d	No; anything could happen in 4 hours it is too far from the end of the graph	B1	
2a	30	B1	
2b	Totals all values in table (= 288)	M1	
	288 ÷ 30	M1	You must include the 12 students who were never late and divide by 30.
	9.6	A1	
3a	94.8p	B1	
3b	2006, price index shot up	B1	
3c	Faster, petrol has gone up by 58%	B1	

117   Scatter diagrams

Q	Answer	Mark	Comment
1a		B2	You will lose a mark if you misplot two values.
1bi	G	B1	
1bii	I	B1	
1c	(See graph for LOBF)	B1	
1d	54 minutes	B1	
2	A and Y; B and W; C and Z; D and X	B2	You will get B1 if you get at least two correct.

118   Cumulative frequency and box plots

Q	Answer	Mark	Comment
1a	$3\frac{1}{4}$ minutes (3 m 15 s)	B1	
1b	2.5 and 3.9 seen	M1	
	1.4 min (1 m 24 s)	A1	
1c	48	B1	
2a	12	B1	
2b	52	B1	
2c	10K higher median; highest score	B1	
2d	10J smaller interquartile range; smaller range	B1	
3	A and X; B and Y; C and W; D and Z	B2	Mark approximate medians and quartiles on the graphs. You will get B1 if you get at least two correct.

119   Probability

Q	Answer	Mark	Comment
1a	Any frequency ÷ 120	M1	
	0.15, 0.06, 0.18, 0.18, 0.29, 0.14	A1	
1b	5; as score 5 had the most frequent result	B1	
2a	Because they cannot happen at the same time	E1	
2b	Because their probabilities add up to 1	E1	
2c	The probability of red is $\frac{2}{5}$	B1	
2d	12	B1	
2e	80	B1	
3	Any combination of two bags and the probability of white calculated. A and B $\frac{6}{13}$ = 0.4615..; A and C $\frac{7}{15}$ = 0.4666; B and C $\frac{7}{16}$ = 0.4375	M1	Show clearly that you are combining two bags. Write 'A and B give 6 white in a total of 13', for example. You will have to use a calculator to work out the fractions as decimals.
	Two combinations worked out	A1	
	A and C stated and all combinations worked out	A1	

120   Probability

Q	Answer	Mark	Comment
1a	22	B1	
1b	PE	B1	
1c	Maths $\frac{5}{12} \approx 42\%$; Science $\frac{7}{18} \approx 39\%$	B1	
1d	$\frac{12}{40} = \frac{3}{10}$	B1	
2a	0.24	B1	
2b	$432 \div 0.36$	M1	
	1200	A1	
2c	$0.24 \times 26\,000$	M1	
	6240	A1	
3	2, 5, 7; 8, 10, 18; 10, 15, 25	B3	Check that your final table satisfies the initial conditions. You will lose a mark for every condition that is not met.

Page 121   Probability

Q	Answer	Mark	Comment
1a	<table><tr><td></td><td></td><td colspan="4">First score</td></tr><tr><td></td><td></td><td>1</td><td>2</td><td>3</td><td>4</td></tr><tr><td></td><td>1</td><td>2</td><td>2</td><td>4</td><td>4</td></tr><tr><td>Second</td><td>2</td><td>2</td><td>4</td><td>6</td><td>8</td></tr><tr><td>score</td><td>3</td><td>4</td><td>6</td><td>6</td><td>12</td></tr><tr><td></td><td>4</td><td>4</td><td>8</td><td>12</td><td>16</td></tr></table>	B2	You will lose a mark for every wrong entry in the table.
1b	1	B1	
1c	$\frac{6}{16} = \frac{3}{8}$	B1	
2a	$\frac{7}{10}; \frac{3}{10}, \frac{7}{10}; \frac{3}{10}, \frac{7}{10}$	B1	
2b	$\frac{3}{10} \times \frac{3}{10}$	M1	
	$\frac{9}{100}$	A1	
2c	$\frac{3}{10} \times \frac{7}{10} + \frac{7}{10} \times \frac{3}{10}$	M1	
	$\frac{42}{100} = \frac{21}{50}$	A1	
3a	$\frac{16}{19}$ not a square number top and bottom	B1	
3b	$\frac{4}{25}$ as this means 2 white balls and 3 black balls	B1	

Page 122   Probability

Q	Answer	Mark	Comment
1a	$\frac{1}{5}$	B1	Be careful as a common wrong answer is $\frac{1}{4}$. There are 4 blue counters for every 1 red counter.
1b	24	B1	
1c	18	B1	
2a	$1 - x$	B1	
2b	x^3	B1	
3a	$\frac{4}{6}, \frac{1}{5}, \frac{4}{5}, \frac{2}{5}, \frac{3}{5}$	B2	Remember that if an egg is used it cannot be used again, so the probabilities change each time. You will lose 1 mark for every wrong entry.
3b	$\frac{2}{6} \times \frac{1}{5} + \frac{4}{6} \times \frac{3}{5}$	M1	
	$\frac{14}{30} = \frac{7}{15}$	A1	
4a	$\frac{x}{x + y} = \frac{y}{x + y - 5}$	M1	
	$x^2 + xy - 5x = xy + y^2$	A1	Getting to this stage will be enough evidence for the examiner as the given answer follows from this.
4b	Any attempt to substitute a value for x and square root the answer	M1	Start with $x = 5$, as otherwise $x^2 - 5x$ will be negative, and increase x by 1 each time.
	$x = 9$ and $y = 6$	A1	This is the only combination that gives a square number answer for $x^2 - 5x$.

124 Number

Q	Answer	Mark	Comment
1a	$34 \times 14 + 12 \times 8 \ (= 476 + 96)$	M1	
	572	A1	
1b	$(700 - 572) \div 20 \ (= 6.4)$	M1	
	14 (7 of each)	A1	
2ai	18	B1	
2aii	24	B1	
2bi	£6.24	B1	
2bii	£300	B1	
3	Testing two values, one twice as big as the other, and subtracting 14	M1	You can also do an algebraic method $x = 2y$; $x - 14 = 4(y - 14)$
	Repeating for an improved pair of values	M1	Substitute one equation into the other to get $2y - 14 = 4y - 56$ or $x - 14 = 2x - 56$
	42	A1	Solve the equations.

125 Number

Q	Answer	Mark	Comment
1ai	70 000	B1	
1aii	0.067	B1	
1bi	At least two numbers rounded to 300, 8 and 0.4	M1	
	6000	A1	
1bii	At least two numbers rounded to 500, 0.5 and 4	M1	
	250	A1	
2ai	370	B1	
2aii	250	B1	
2bi	0.76	B1	
2bii	0.0065	B1	
2ci	12 000 000	B1	
2cii	360 000	B1	
2di	3 000	B1	
2dii	500	B1	
3a	Both round to 3.7	B1	
3b	Both have 4 sf	B1	
3c	$3.65 \div 10$	B1	

126 Number

Q	Answer	Mark	Comment
1a	$p = 2$, $q = 3$	B2	Values can be either way round. 1 mark for each.
1b	Any indication that 360 split into a product including at least one prime number	M1	
	$2^3 \times 3^2 \times 5$	A1	
1c	$a = 2$, $b = 7$	B2	B1 for each correct answer.
1d	$2^2 \times 7^2$	B1	
2a	$2^3 \times 3$	B1	
2b	$2^2 \times 3 \times 5$	B1	
2c	120	B1	
2d	12	B1	
2ei	$2^4 \times 3^2 \times 5^2$	B1	
2eii	$2^2 \times 3 \times 5$	B1	
3a	$a^2 b^2 c^3$	B1	Take the highest power of each letter.
3b	abc	B1	Take the lowest power of each letter.

127 Fractions

Q	Answer	Mark	Comment
1a	$\frac{15}{20} + \frac{8}{20}$	M1	
	$1\frac{3}{20}$	A1	
1b	$\frac{11}{3} - \frac{9}{5}$	M1	
	$\frac{55}{15} - \frac{27}{15}$	M1	
	$1\frac{13}{15}$	A1	

Workbook answers

1c	$1 - \frac{2}{5} - \frac{1}{4} - \frac{1}{6}$	M1	
	$1 - \frac{24}{60} - \frac{15}{60} - \frac{10}{60}$	M1	
	$\frac{11}{60}$	A1	
2a	$3\frac{1}{2}$	B1	
2b	$1\frac{3}{8}$	B1	
2c	$\frac{1}{2} \times \frac{9}{10} \times \frac{13}{6}$	M1	
	$\frac{39}{40}$	A1	
3	$\frac{3}{2}$ is the odd one out because it is the only top-heavy fraction	E1	
	$\frac{2}{3}$ is the odd one out because it is the only one that is a recurring decimal	E1	
	$\frac{2}{8}$ is the odd one out as it is the only one that is not a vulgar fraction (i.e. cancelled down)	E1	

128 Percentage

Q	Answer	Mark	Comment
1a	$6000 \times 0.88 \times 0.9$	B1	
1b	6.80, 3.40 and 1.70 seen	M1	
	£11.90	A1	
2a	700×1.175	M1	The multiplier for an increase of $17\frac{1}{2}$% is 1.175.
	£822.50	A1	Remember to give money answers to 2 decimal places. An answer of £822.5 would get A0.
2b	250×0.88	M1	A reduction of 12% is a multiplier of 0.88.
	£220	A1	
3	$2655 - 280 = 405$	M1	
	$405 \div 2250 \ (\times 100)$	M1	Remember to divide the increase by the original amount.
	18%	A1	
4	$1.1 \times 0.9 \ (= 0.99)$	M1	You can choose a value to start with, say £100, and work out a 10% increase (£110) and then a 10% decrease of this (£99).
	Explanation that this is 99% of the original	A1	

129 Percentage

Q	Answer	Mark	Comment
1a	$8000 \times 0.88 \times 0.88$	B1	
1b	Sight of 1.035	B1	
	2000×1.035^6	M1	
	£2458.51	A1	Round off the answer to 2 decimal places as it a money answer.
1c	Sight of 0.98	B1	
	2000×0.98^6	M1	
	£1771.68	A1	
2a	240×0.95 or 260×0.88	M1	
	288 or 228.80	A1	
	Electro Co and both values seen	A1	
2b	Sight of 0.95	B1	
	$361 \div 0.95$	M1	
	£380	A1	
3	Sight of 1.05 or 0.97	B1	
	Both same and $1.05 \times 0.97 = 0.97 \times 1.05$ seen	E2	You can choose a value, say £100, and do a 5% increase followed by a 3% decrease (£101.85) and show that this gives the same answer as a 3% decrease followed by a 5% increase.

130 Ratio

Q	Answer	Mark	Comment
1a	4 : 3	B1	
1b	1 : 0.4	B1	

1c	$1000 \div 8 \ (\times 3)$	M1	
	375 ml	A1	
2	$15 \div 3 \ (\times 7)$	M1	
	35	A1	
3a	$72 \div 2.25$	M1	
	32	A1	
	km/h	B1	
3b	9 km seen	B1	Mark is for numerical answer only.
	$63 \div 1.75$	M1	
	36 km/h	A1	
4	C, F, F, T	B3	You will lose 1 mark for each wrong answer.

131 Ratio

Q	Answer	Mark	Comment
1a	$50 \div 275 \times 165$	M1	
	30 litres	A1	
1b	$275 \div 50 \times 26$	M1	
	143 miles	A1	
2	$600 \div 155$ or $800 \div 220$	M1	
	3.87 or 3.63	A1	
	Handy size and both values seen	A1	
3	$5.1 \div 750$	M1	
	Digits 68 seen	A1	
	6.8 g/cm³	A1	
4	2 hours $+ \frac{1}{2}$ hour	B1	
	$40 \div 2.5$	M1	
	No and 16 km/h seen	A1	

132 Powers and reciprocals

Q	Answer	Mark	Comment
1ai	13	B1	
1aii	125	B1	
1bi	4	B1	
1bii	256	B1	
2a	64, 4; 256, 6; 1024, 4	B2	You will lose 1 mark for every wrong answer.
2b	4, all odd powers end in 4	B1	
2c	$5^6 = 15625 > 6^5 = 7776$	B1	
3a	$\frac{1}{8}$	B1	
3b	$\frac{1}{16}$	B1	
3c	$\frac{1}{8} + \frac{1}{4}$	M1	
	$\frac{3}{8}$	A1	
4	2 of $\frac{1}{8}, \frac{1}{9}, \frac{1}{16}$ seen	M1	
	2^{-4}, 3^{-2}, 8^{-1} and all values seen	B1	

133 Powers and reciprocals

Q	Answer	Mark	Comment
1a	x^7	B1	
1b	x^4	B1	
1ci	4	B1	
1cii	3	B1	
1d	1 000 000	B1	
2a	$16a^4b^3$	B2	You will lose 1 mark for each error.
2b	$9x^4y^6$	B2	You will lose 1 mark for each error.
3a	2	B1	
3b	$\frac{1}{11}$	B1	
3c	$\frac{1}{27}$	B2	
	$\sqrt[4]{81} = 3$	B1	

Workbook answers

4	2 of $\frac{1}{9}$, $\frac{1}{8}$, $\frac{1}{4}$ seen	M1	
	$27^{-\frac{2}{3}}$ 2^{-3} $(\sqrt{16})^{-1}$ and all values seen	B1	

134 Powers and reciprocals

Q	Answer	Mark	Comment
1a	4.52×10^4	B1	
1b	0.006	B1	
1c	1.8×10^6	B1	
1d	8×10^{-4}	B1	
1e	2×10^2	B1	
1f	6×10^2	B1	
2ai	0.4444...	B1	
2aii	0.5555...	B1	
2bi	$\frac{1}{10}$	B1	
2bii	$1\frac{1}{3}$	B1	
2ci	0.1	B1	
2cii	$1.\dot{3}$	B1	
2di	0.8	B1	
2dii	0.4	B1	
2diii	0.2	B1	
3	Substituting n = 3, 4, 5, and 6 into the formula	M1	
	$\frac{9}{4}$, $\frac{12}{5}$, $\frac{15}{6}$, $\frac{18}{7}$	A1	
	2.25, 2.4, 2.5, 2.571...	A1	

135 Surds

Q	Answer	Mark	Comment
1a	Splitting $\sqrt{48}$ into two with one square value e.g. $\sqrt{(4 \times 12)}$, $\sqrt{(16 \times 3)}$	M1	
	$4\sqrt{3}$	A1	
1b	$\sqrt{(16 \times 3)} + \sqrt{(25 \times 3)}$	M1	
	$9\sqrt{3}$	A1	
1c	$3 + 5\sqrt{3} - \sqrt{3} - 5$	M1	
	$-2 + 4\sqrt{3}$	A1	
1d	$5 + 2\sqrt{5} - 2\sqrt{5} - 4$	M1	
	1	A1	
2a	$\frac{3\sqrt{6}}{\sqrt{6}\sqrt{6}}$	M1	
	$\frac{\sqrt{6}}{2}$	A1	
2b	$20 \div 5\sqrt{2}$	M1	
	$2\sqrt{2}$	A1	
2c	$\frac{16 + 12}{4\sqrt{12}}$	M1	
	$\frac{28\sqrt{12}}{48}$	A1	It is sufficient to get to this stage as the next step is the answer which is given.
3	$ac \div bd$ must be a square number	B1	

136 Variation

Q	Answer	Mark	Comment
1a	$m = ks^3$	B1	
1b	$50 = k(10)^3$	M1	
	$\frac{1}{20}$	A1	
1c	$\frac{1}{20} \times (20)^3$	M1	
	400	A1	
1d	$3.2 = \frac{1}{20} \times s^3$	M1	
	4	A1	
2ai	$2 = k \times 4^2$	M1	
	$\frac{1}{8}$	A1	

2aii	$2 = \frac{k}{\sqrt{4}}$	M1	
	4	A1	
2b	$y = \frac{k}{3\sqrt{x}}$	M1	
	$k = 20$	A1	
	4	A1	
3	$y = \frac{k_1}{\sqrt{x}}$	M1	
	$z = k_2 y^3$	M1	
	$k_1 = 6$	A1	
	$k_2 = \frac{1}{3}$	A1	
	576	A1	

137 Limits

Q	Answer	Mark	Comment
1ai	45	B1	
1aii	54	B1	
1bi	45	B1	
1bii	55	B1	
2a	$(19.5)^3$	M1	
	7414.875	A1	Never round off limits answers to less than 4 sf.
2b	$(20.5)^3$	M1	
	8615.125	A1	
3	225 to 235 and 395 to 405	M1	Be careful with the upper limit. It is always the same difference from the original value as the lower limit.
	$(235) \div (395)^2$	M1	
	0.001 506	A1	
4	395 and 405	B1	
	$\sqrt[3]{\frac{3 \times 395}{4\pi}} = 4.5517$ or $\sqrt[3]{\frac{3 \times 405}{4\pi}} = 4.5897$	M1	
	$4\pi \times$ limit^2	M1	
	260.3 to 264.7	A1	

139 Circles and area

Q	Answer	Mark	Comment
1a	$\pi \times 12 \times 12$	M1	You will need to learn the formula for the area of a circle.
	452.4	A1	
1b	$\pi \times 20$	M1	You will need to learn the formula for the perimeter of a circle.
	62.8	A1	
2	$0.5 \times \pi \times 10 \times 10$	M1	
	50π	A1	
3	$0.5 \times 8 \times (16 + 22)$	M1	The area of a trapezium is given on the formula sheet.
	152	A1	
4a	$4x = x^2$	M1	Equate the area and the volume.
	4	A1	
4b	$2\pi \times r = \pi r^2$	M1	Cancel out one r and π.
	2	A1	

140 Sectors and prisms

Q	Answer	Mark	Comment
1a	$72 \div 360 \times \pi \times 100$	M1	You will need to recall the formula for the area of a sector.
	20π	A1	
1b	$288 \div 360 \times 2 \times 10 \times \pi + 20$	M1	You will need to recall the formula for the arc length of a sector. Remember to add on the two radii.
	$20 + 16\pi$	A1	
2a	$0.5 \times 7 \times 4$	M1	
	14	A1	
2b	14×12	M1	This mark would follow through from your answer in **a**.
	168	A1	
3	$\theta \div 360 \times 2 \times \pi \times r = r$	M1	Cancel r.
	$\theta = 360 \div 2\pi$	A1	
	57.3	A1	

Workbook answers

141 Cylinders and pyramids

Q	Answer	Mark	Comment
1a	$\pi \times 4 \times 4 \times 10$	M1	You will need to learn the formula for the volume of a cylinder: $V = \pi r^2 h$.
	160π	A1	
1b	$2 \times \pi \times 4 \times 10 + 2 \times \pi \times 4^2$	M1	The total surface area is the curved surface area plus two circular ends: $2\pi rh + 2\pi r^2$.
	112π	A1	
2	$\frac{1}{3} \times 10 \times 8 \times 12$ or $\frac{1}{3} \times 2.5 \times 2 \times 3$	M1	Substitute the values for the volume of one of the pyramids to get a method mark.
	$\frac{1}{3} \times 10 \times 8 \times 12 - \frac{1}{3} \times 2.5 \times 2 \times 3$	M1	Subtract the two pyramid volumes.
	315	A1	
3	Side of square base $= x$ $2x^2 = (2r)^2$	M1	Use Pythagoras to work out the length of the side of the square base in terms of r.
	$x = \sqrt{2} \times r$	A1	
	Volume pyramid $= \frac{1}{3} \times \sqrt{2} \times r \times \sqrt{2} \times r \times h$	M1	
	$\frac{2hr^2}{3} \div \pi r^2 h \times 100$	M1	Cancel the hr^2.
	$= 21.22\%$	A1	Show the exact value, which obviously rounds off to 21%.

142 Cones and spheres

Q	Answer	Mark	Comment
1	$\frac{1}{3} \pi \times 0.5^2 \times 6$ (or 3)	M1	Give the volume of either cone for a method mark.
	$\frac{1}{3} \pi \times 0.5^2 \times 6 + \frac{1}{3} \pi \times 0.5^2 \times 3$	M1	Add the two cones together.
	2.356 cm^3	A1	
	2.356×1.5	M1	Mass = volume × density gets a method mark.
	3.5	A1	
2	$\pi \times 6^2 \times 6 \ (= 678.6)$	M1	Volume of the cylinder gets a method mark.
	$0.5 \times \frac{4}{3} \times \pi \times 5^3 \ (= 261.8)$	M1	Volume of the hemisphere gets a method mark.
	$687.6 - 261.8$	M1	Subtract the volumes.
	416.8	A1	
3	$\frac{4}{3}\pi r^3 = \frac{1}{3}\pi\left(\frac{1}{2}r\right)^2 h$	M1	Equate the volume of the cone and the sphere. Remember the radius of the cone is $\frac{1}{2}r$.
	$\frac{4}{3}\pi r^3 = \frac{1}{12}\pi r^2 h$	M1	Cancel π and r^2.
	$(h =) \ 16r$	A1	

143 Pythagoras' theorem

Q	Answer	Mark	Comment
1	$10^2 + 8^2$	M1	
	$\sqrt{164}$	M1	
	12.8	A1	
2	$15^2 - 11^2$	M1	
	$\sqrt{104}$	M1	
	10.2	A1	
3	$2.1^2 + 4^2$	M1	
	$\sqrt{20.41}$	M1	
	4.5	A1	
4	First hypotenuse$^2 = 1^2 + 1^2$	M1	
	so $= \sqrt{2}$	A1	
	Second hypotenuse$^2 = (\sqrt{2})^2 + 1^2$ so $= \sqrt{3}$	M1	Spot the pattern that the hypotenuse are going up $\sqrt{2}$, $\sqrt{3}$, $\sqrt{4}$, etc.
	3	A1	This is $\sqrt{9}$.

144 Pythagoras' theorem

Q	Answer	Mark	Comment
1	$\sqrt{(12^2 - 3.5^2)}$	M1	You will get a mark for drawing the vertical bisector on the triangle and using Pythagoras on the remaining triangle.
	$\sqrt{131.75}$	M1	
	3.5×11.48	M1	
	40.2	A1	

2a	$x^2 = \sqrt{200}$	M1	The length AC is the diagonal of a square with sides 10 cm.
	14.14	A1	
2b	$\sqrt{(15^2 - 7.07^2)}$	M1	Half the length AC is AX
	13.2	A1	
2c	$\sqrt{(15^2 - 5^2)}$	M1	EAB is an isosceles triangle.
	14.14	A1	
3a	$AM^2 = 3^2 - 1^2$	M1	The distance from M to the end of the roof is 1 m. The hypotenuse is 3 m.
	$AM = \sqrt{8}$	A1	
	$AX^2 = (\sqrt{8})^2 - 1.5^2$	M1	
	$AX = 2.4$	A1	
3b	Volume prism = $1.5 \times 2.4 \times 4$	M1	The prism has an isosceles triangle end with a base of 3 and a height AX.
	= 14.4 m³	A1	
	Volume pyramid = $\frac{1}{3} \times 3 \times 2 \times 2.4$ (= 4.8 m³)	M1	The pyramid has a rectangular base, 2 by 3, and a height AX.
	Total = 19.2	A1	

145 Using Pythagoras and trigonometry

Q	Answer	Mark	Comment
1a	0.5449883506	B1	
1b	0.54 or 0.545	B1	
2a	$13 \times \sin 40$	M1	
	8.36	A1	
2b	$\sin^{-1}(8 \div 15)$	M1	
	32.2	A1	
3a	$15 \div \cos 48$	M1	
	22.4	A1	
3b	$\cos^{-1}(9 \div 20)$	M1	
	63.2	A1	
4a	$10 \times \tan 20$	M1	
	3.64	A1	
4b	$\tan^{-1}(3 \div 5)$	M1	
	31.0	A1	
5a	$\sin A = \frac{a}{c}$	E1	
	$\cos B = \frac{a}{c}$	E1	It is enough to show that these ratios are the same fraction.
5b	$\tan A = \frac{a}{b}$ $\tan B = \frac{b}{a}$	E1	
	$\frac{a}{b} = \frac{1}{\frac{b}{a}}$	E1	Dividing by a fraction is just the reciprocal of the fraction.

146 Using Pythagoras and trigonometry

Q	Answer	Mark	Comment
1	Sight of cosine	M1	A method mark is obtained simply by identifying the correct ratio.
	$\cos^{-1}(20 \div 28)$	M1	
	44.4	A1	
2	Sight of tan	M1	A method mark is obtained simply by identifying the correct ratio.
	$12 \div \tan 25$	M1	
	25.7	A1	
3	Sight of sine	M1	A method mark is obtained simply by identifying the correct ratio.
	$20 \times \sin 36$	M1	
	11.8	A1	
4	Sight of sine	M1	A method mark is obtained simply by identifying the correct ratio.
	$\sin^{-1}(40 \div 50)$	M1	
	53.1	A1	
5	Total height = $2 \times \sin 60 + 1.5 \times \sin 20$	M1	Split the roof into right-angled triangles with hypotenuse of 2 and 1.5 and angles of 60 and 20.
	Total horizontal = $2 \times \cos 60 + 1.5 \times \cos 20$	M1	Work out the total distances BC and AB.
	2.25 and 2.41	A1	
	$\tan^{-1}(2.25 \div 2.41)$	M1	
	43.0	A1	

147 More advanced trigonometry

Q	Answer	Mark	Comment
1a	Sight of tan	M1	A method mark is obtained simply by identifying the correct ratio.
	$15 \div \tan 68$	M1	
	6.1	A1	
1b	$15 \div \tan 42$	M1	
	16.7	A1	Calculate the length AC.
	10.6	A1	Subtract BC from AC.
2a	$8 \times \sin 65$ or $8 \times \cos 25$	M1	
	2×7.25	M1	
	14.5	A1	
2b	65	B1	
2c	Sight of sine with angle 32.5 or cosine with angle 57.5	M1	
	$7.25 \div \sin 32.5$ or $7.25 \div \cos 57.5$	M1	
	13.5	A1	
3	$AD = 10 \tan 28$	M1	Work out AD.
	5.32	A1	
	$AC = 17.32$	A1	Work out AC.
	$\tan^{-1}(17.32 \div 10)$	M1	Use tangent to find the angle.
	60.0	A1	

148 3-D trigonometry

Q	Answer	Mark	Comment
1	$\sqrt{(6^2 + 15^2)}$	M1	Identify a triangle that contains the angle, say BDF and find a missing side, BD.
	16.16	A1	
	$\tan^{-1}(8 \div 16.16)$	M1	Use tangent to find the angle.
	26.3	A1	
2	$12^2 - 4^2$ or $8^2 - 4^2$	M1	Find the sides of the right-angled triangle CMD.
	11.3	A1	
	6.93	A1	
	$\tan^{-1}(6.93 \div 11.3)$	M1	
	31.5	A1	
3	$5 \div \tan 30$	M1	Find the length AC. (AC = EG)
	8.66	A1	
	$\sqrt{(8.66^2 - 8^2)} = 3.32$	M1	Find the length HG.
	$\sqrt{(3.32^2 + 3^2)} = 4.47$	M1	Find the length AB.
	$\tan^{-1}(5 \div 4.47)$	M1	Use tangent to find the angle.
	48.2	A1	

149 Trigonometric ratios of angles from 0° to 360°

Q	Answer	Mark	Comment
1a	216.9° and 323.1°	B2	Draw a line from 0.6 across to the graph and read off the values. A small degree of tolerance is allowed.
1b	188.6° and 351.4°	B2	
2a	41.4° and 318.6°	B2	A small degree of tolerance is allowed.
2b	104.5° and 255.5°	B2	
3a	$\sin 142 = 0.616$; $\sin 218 = -0.616$; $\sin 322 = -0.616$	B2	You will lose 1 mark for each error.
3b	$\cos 52 = 0.616$; $\cos 128 = -0.616$; $\cos 232 = -0.616$; $\cos 308 = 0.616$	B2	You will lose 1 mark for each error.

150 Sine rule

Q	Answer	Mark	Comment
1	$\dfrac{\sin B}{20} = \dfrac{\sin 78}{25}$	M1	Use the sine rule with the angles on top.
	$\sin B = \dfrac{20 \times \sin 78}{25}$	M1	
	51.5	A1	
2a	$\dfrac{x}{\sin 68} = \dfrac{15}{\sin 55}$	M1	
	$x = \dfrac{15 \times \sin 68}{\sin 55}$	M1	
	17.0	A1	

2b	$\frac{\sin B}{15} = \frac{\sin 32}{10}$	M1	
	$\sin B = \frac{15 \times \sin 32}{10}$	M1	
	127.4	A1	Remember to take the answer away from 180 to get the obtuse angle.
3	$\sin x = \frac{6 \times \sin 40}{3}$	M1	
	$\sin x = 1.2855$	A1	
	Sin x cannot be bigger than 1, but this triangle cannot be made as the side of 3 cm would not reach the side of 6 cm.	E1	If you try to construct this triangle it is not possible – which is what the mathematics shows you by giving an impossible value.

151 Cosine rule

Q	Answer	Mark	Comment
1	$\cos C = \frac{21^2 + 22^2 - 28^2}{2 \times 21 \times 22}$	M1	You will have to learn the rule for the angle as this is not given on the formula sheet.
	0.152 597	A1	
	81.2	A1	
2a	$25^2 + 29^2 - 2 \times 25 \times 29 \times \cos 58$	M1	
	697.62	A1	Make sure you can work this out with your calculator.
	26.4	A1	
2b	$\cos B = \frac{10^2 + 17^2 - 22^2}{2 \times 10 \times 17}$	M1	
	−0.2794	A1	
	106.2	A1	The negative value will give the obtuse angle straight away with a calculator.
3	$\cos x = \frac{7^2 + 8^2 - 15^2}{2 \times 7 \times 8}$	M1	
	$\cos x = -1$	A1	
	$x = 180$ as the sides 7 and 8 would be flat against the side 15, making a straight line.	A1	The mathematics proves that these three sides make a straight line.

152 Solving triangles

Q	Answer	Mark	Comment
1a	$9^2 + 5^2 - 2 \times 5 \times 9 \times \cos 110$	M1	
	136.8	A1	
	11.7	A1	
1b	$\frac{\sin x}{10} = \frac{\sin 88}{11.7}$	M1	
	$\sin x = \frac{10 \times \sin 88}{11.7}$	M1	
	58.7	A1	
2a	$0.5 \times 8 \times 9 \times \sin 40$	M1	
	23.1	A1	
2b	$\frac{\sin R}{7} = \frac{\sin 106}{12}$	M1	
	$R = 34.1, P = 39.9$	A1	
	$0.5 \times 7 \times 12 \times \sin 39.9$	M1	
	26.9	A1	
3	$9^2 + 10^2 - 2 \times 9 \times 10 \times \cos 70$	M1	Find the length BD.
	10.93	A1	
	$0.5 \times 10 \times 9 \times \sin 70$	M1	Find the area of triangle BCD.
	42.29	A1	
	Area ABD = 17.71	M1	
	$0.5 \times 6 \times 10.93 \times \sin ADB = 17.71$	M1	Find the angle ADB by using the area formula.
	32.7°	A1	
	$6^2 + 10.93^2 - 2 \times 6 \times 10.93 \times \cos 32.7$	M1	Use the cosine rule to find AB.
	6.72	A1	

153 Polygons

Q	Answer	Mark	Comment
1	BCD = 100° (angles in a kite)	B1	If the question asks for reasons, give them or you will lose marks.
	BCF = 120° (angles in a parallelogram)	B1	
	DCE = 80° (angles on a straight line)	B1	
	CED = 40° (alternate to ECF)	B1	
2a	Draw the diagonals from one vertex.	M1	
	This gives three triangles 3 × 180 = 540	A1	

—Workbook answers—

2b	360 ÷ 36	M1	
	10	A1	
2c	Exterior angle = 20	M1	
	360 ÷ 20 = 18	A1	
3a	All angles 90°	B1	
3b	Diagonals cross at right angles	B1	
3c	Rotational symmetry of order 2	B1	Other answers are possible, e.g. oppostie sides are equal, diagonals bisect each other.
3d	Any valid answer for pair chosen e.g. rhombus and square have all sides equal	B1	When you are given a choice, go for something straightforward.

154 Circle theorems

Q	Answer	Mark	Comment
1a	27	B1	
	ACB = 90°, angles in a triangle	B1	
1b	Angles in an isosceles triangle	B1	
2a	102	B1	
	Opposite angles in a cyclic quadrilateral	B1	
2b	156	B1	
	Angle at centre is twice the angle on the circumference	B1	
3a	$\sqrt{(8^2 + 1^2)}$	M1	As ABC is a right angle, AC must be a diameter.
	$\sqrt{65}$	A1	If the question asks for an exact answer, just leave it as a square root.
3b	Point marked halfway between A and C	B1	

155 Circle theorems

Qu	Answer	Mark	Comment
1a	33	B1	
	AQB is a right angle, angles in a triangle	B1	
1b	33	B1	
	Alternate segment theorem	B1	
2	ACB = x (isosceles)	B1	
	BAP = CAT = x (alternate segment)	B1	
	BAC is symmetric so AM must be at right angles to PT	B1	
3	ABC = 65 (angle in an isosceles triangle)	B1	There are alternative acceptable methods of finding BDO.
	ABD = 155 (CBD = 90)	B1	
	ACD = 90	B1	
	BDO = 65 (angles in a quadrilateral)	B1	

156 Transformation geometry

Q	Answer	Mark	Comment
1	R	B1	
	S	B1	
2a	Translation	B1	
	$\binom{-6}{-3}$	B1	
2b	Triangle at (3, 6), (3, 9), (5, 6)	B1	
2c	$\binom{-7}{5}$	B1	
3		B3	You will get 1 mark for a sketch of any triangle with sides 7 cm and 8 cm and an angle of 40°. You will get 1 mark for a different triangle and 1 mark if they are not congruent.

157 Transformation geometry

Q	Answer	Mark	Comment
1		B3	B1 for each correct answer.

2ab		B2	B1 for each correct answer.
2c	90°	B1	
	Clockwise	B1	
	(4, 3)	B1	
3	Translation of $\binom{3}{-3}$	B1	
	Reflection in $y = x$	B1	
	Rotation of 180°	B1	There are other rotations, such as 90° anticlockwise about (1, 1).
	Centre (3.5, 3.5)	B1	

158 Transformation geometry

Q	Answer	Mark	Comment
1a	Enlargement scale factor $\frac{1}{2}$	B1	
	Centre (1, 8)	B1	
1b	Rays from (1, 0) to vertices of triangle	M1	
	Triangle at (2, 1), (3, 1), (2, $2\frac{1}{2}$)	A1	
2ab		B2	B1 for each correct trtiangle.
2c	Reflection in $y = -x$	B1	
3	Triangle A at (4, 8), (5,8), (4, 10)	B1	
	Triangle B at (8, 4), (8, 8), (10, 4)	B1	
	Enlargement sf $1\frac{1}{2}$ about (4, 0)	B1	

159 Constructions

Q	Answer	Mark	Comment
1	Side of 7 cm	B1	You are allowed to be up to 1 mm out.
	Side of 5 cm	B1	
	Angle of 55°	B1	You are allowed to be 1° out.
2a	Arcs shown	M1	
	Angle 60° ± 1°	A1	
2bi	Side of 8 cm	B1	
	Side of 6 cm	B1	
	Angle of 60°	B1	

2bii	7.2 cm	B1	
3	Construct an angle of 60°	B1	
	Bisect the 60° angle	B1	

160 Constructions

Q	Answer	Mark	Comment
1a	Arcs either side	M1	
	Bisector drawn within 1 mm of centre and at 90°	A1	
1b	Arcs shown	M1	
	Angle 90° ± 1°	A1	
2a	Arcs either side	M1	
	Bisector drawn within 1 mm of centre and at 90°	A1	
2b	Arcs shown	M1	
	Bisector within ±1°	A1	
3	Bisect the line	B1	
	Then bisect each of the halves	B1	

161 Constructions and loci

Q	Answer	Mark	Comment
1	Arcs from C on line	M1	
	Arcs bisecting on other side	M1	
	Line 90° ± 1°	A1	
2a	Circle radius 4 cm around flower bed	B1	
2b	Rectangle drawn 1 cm from edge	B1	
	Overlapping area shaded	B1	
3	Distance from centre to each side is radius r. Area of A for example is $\frac{1}{2}ar$, so ratio area is $\frac{1}{2}ar : \frac{1}{2}br : \frac{1}{2}cr$. Cancel the $\frac{1}{2}r$ and it equals $a : b : c$	E3	E2 for showing area ratios. E1 for distance from centre to edge is r.

162 Similarity

Q	Answer	Mark	Comment
1a	sf = 9.6 ÷ 6.4 = 1.5	M1	Work out a scale factor using corresponding distances on the diagram.
	8.4	A1	
1b	8.8	B1	
2	1500 ÷ 20 = 75	M1	
	width = 6	A1	
	length = 9	A1	
3	5 ÷ 0.4 × 0.8 = 10	M1	Alternative methods are aceptable.
	0.6 × 2 = 1.2	A1	
	1.2 + 1.2 + 0.6	M1	
	3 m	A1	

163 Similarity

Q	Answer	Mark	Comment
1a	3.6 or 2.4 × 0.75	M1	Multiply by the linear scale factor.
	2.7 by 1.8	A1	
1b	52 × (0.75)²	M1	Multiply by the area scale factor.
	29.25 m²	A1	
1c	75 × (0.75)³	M1	Multiply by the volume scale factor.
	31.6 cm³	A1	
2a	21 × ³√2.5	M1	Volume scale factor is 2.5 so linear sf is ³√2.5.
	28.5 cm	A1	
2b	70 × (³√2.5)²	M1	
	38 cm²	A1	
3	9 ÷ 5 = 1.8	M1	
	1.8³ = 5.832	A1	
	Claim is justified as 5.8 is about 6 or Not justified as 5.8 is less than 6	A1	

164 Vectors

Q	Answer	Mark	Comment
1a		B4	B1 for each correct answer.
1b	$2\frac{1}{2}\mathbf{a} + \mathbf{b}$	B1	
1c	$5\mathbf{a} - 7\mathbf{b}$	B1	
2a	$q_1 = \binom{2}{3}$ and $q_2 = \binom{3}{-2}$	B1	
	$r_1 = \binom{2}{5}$ and $r_2 = \binom{5}{-2}$	B1	
	$s_1 = \binom{1}{1}$ and $s_2 = \binom{1}{-1}$	B1	
2b	$ac + bd = 0$	B1	

165 Vectors

Q	Answer	Mark	Comment
1ai	$-\frac{1}{2}\mathbf{a} + \frac{1}{2}\mathbf{c}$	B1	
1aii	$\mathbf{b} - \mathbf{a}$	B1	
1aiii	$\mathbf{c} - \mathbf{b}$	B1	
1aiv	$-\frac{1}{2}\mathbf{a} + \frac{1}{2}\mathbf{c}$	B1	
1b	Parallelogram as PS and QR are equal so parallel and same length	B1	

2	$\overrightarrow{MN} = -\frac{4}{5}\mathbf{a} + \frac{2}{3}(\mathbf{a} + \mathbf{b})$	M1	
	$\overrightarrow{MN} = -\frac{2}{15}\mathbf{a} + \frac{2}{3}\mathbf{b}$	A1	
	$\overrightarrow{NX} = -\frac{2}{3}(\mathbf{a} + \mathbf{b}) + \mathbf{b} + n\mathbf{a}$	M1	
	$\overrightarrow{NX} = (-\frac{2}{3} + n)\mathbf{a} + \frac{1}{3}\mathbf{b}$	A1	
	Coefficient of $\mathbf{a}$ is $-\frac{1}{5}$ coefficient of $\mathbf{b}$ so $-\frac{2}{3} + n = -\frac{1}{15}$	M1	
	$3 : 2$	A1	

168   Basic algebra

Q	Answer	Mark	Comment
1a	$2(3)^2 - 3(-2)$	M1	Substituting the numbers into the formula always gets a mark.
	24	A1	
1b	$((-2)^2 - (-6)^2) \div -8$	M1	Remember that $(-2)^2 = +4$ and the $- \div - = +$.
	4	A1	
2a	$5x - 15$	B1	
2b	$2x + 2 + 6x + 4$	M1	
	$8x + 6$	A1	
2c	$3x - 12 + 8x + 2$	M1	
	$11x - 10$	A1	
2d	$2(2x + 3) + 2(x + 3)$	M1	
	$6x + 12$	A1	
3a	$2(2x + 3)$	B1	
3b	$x(5x + 2)$	B1	
4a	Values that work, e.g. $a = 3$, $b = 2$, $21 - 6 = 15$	B1	There are many choices but make sure that you show your answers work.
4b	Values that work, e.g. $a = 2$, $b = 1$, $14 - 3 = 11$	B1	
4c	Values that work, e.g. $a = 2$, $b = 4$, $14 - 12 = 2$	B1	

169   Linear equations

Q	Answer	Mark	Comment
1a	$3x = 1 - 4(-3)$	M1	
	-1	A1	
1b	$7x - 2 = 12$	M1	
	$7x = 14$	A1	
	2	A1	
1c	$4x - 4 = 6$	M1	Alternative methods are acceptable.
	$4x = 10$	A1	
	2.5	A1	
2a	$12y - 8 = 16$	M1	Alternative methods are acceptable.
	$12y = 24$	A1	
	2	A1	
2b	$5x - x = 10 + 2$	M1	
	$4x = 12$	A1	
	$x = 3$	A1	
2c	$5x - 3x = 1 + 2$	M1	
	$2x = 3$	A1	
	1.5	A1	
3	$4(3x + 1) - 2(2x - 1)$	M1	
	$8x + 6 = 16$	M1	
	$x = 1.25$	A1	

170   Linear equations

Q	Answer	Mark	Comment
1a	$3x + 12 = x - 5$	M1	
	$2x = -17$	A1	
	-8.5	A1	
1b	$5x - 10 = 2x + 8$	M1	
	$3x = 18$	A1	
	6	A1	

2	$x + 3x - 1 + 2x + 5 = 25$	M1	
	$6x + 4 = 25$	A1	
	3.5 cm	A1	
3	$8x + 8$ or $11x + 9$	M1	
	$10x = 9x + 17$	M1	
	17	A1	

171 Trial and improvement

Q	Answer	Mark	Comment
1	(5.6 =) 198.016	M1	
	(5.7 =) 207.993	M1	
	Testing 5.65 (202.96) and stating 5.7 as the answer.	A1	
2	Testing 4 (= 8.25)	M1	
	(3.7 =) 7.94	M1	
	(3.8 =) 8.12	M1	
	Testing 3.75 (8.033) and stating 3.7 as answer	A1	
3	$x^2(x + 6) = 2000$	M1	
	(10.8 =) 1959.552	M1	
	(10.9 =) 2007.889	A1	
	Testing 10.85 (1983.624) and stating 10.9 as answer	A1	

172 Simultaneous equations

Q	Answer	Mark	Comment
1a	$9x - 3y = 27$	M1	
	$x = 2.5$	A1	
	$y = -1.5$	A1	
1b	$8x = 10$	M1	
	$x = 1.25$	A1	
	$y = 0.5$	A1	
1c	$15x + 9y = 18$	M1	
	$15x - 35y = 95$	M1	
	$x = 2.25$	A1	
	$y = -1.75$	A1	
2a	$y = 5 - 3x$	B1	
2b	$2x = 22 - 3(5 - 3x)$	M1	
	$2x = 22 - 15 + 9x$	A1	
	$x = -1$ and $y = 8$	A1	
3	$3x + 5y = 375$ or $2x + y = 110$	M1	
	$10x + 5y = 550$	M1	
	$x = 25, y = 60$	A1	
	No as total weight is 510 kg > 500 kg	A1	

173 Simultaneous equations and formulae

Q	Answer	Mark	Comment
1a	$24x + 20y = 134$	B1	
	$20x + 24y = 130$	B1	
1b	$144x + 120y = 804$ or $100x + 120y = 650$	M1	
	$44x = 154$	A1	
	3.5 g	A1	
2a	$x = C/\pi$	B1	
2b	$3x = 6y + 9$	M1	
	$x = 2y + 3$	A1	
2c	$3q - 8 = -2x$	M1	
	$x = (8 - 3q)/2$	A1	
2d	$x = (4y + 2)/2$	M1	
	$x = 2y + 1$	A1	
2e	$4y + 12 = 2x + 2$	M1	
	$2y + 6 = x + 1$	M1	
	$x = 2y + 5$	A1	
3a	$6x = 10y - 124$	M1	
	$y = \dfrac{6x + 124}{10}$	A1	

Q	Answer	Mark	Comment
3b	$y = \dfrac{6 \times 21 + 124}{10}$	M1	
	25p	A1	

174   Algebra 2

Q	Answer	Mark	Comment
1a	$3x + 6$	B1	
1b	$x^2 + 2x$	B1	
1c	$x^2 - 3x + 2x - 6$	M1	
	$x^2 - x - 6$	A1	
1d	$(x + 2)(x + 1)$	M1	
	$x^2 + 3x + 2$	A1	
2a	$x^2 - 4x + x - 4$	M1	
	$x^2 - 3x - 4$	A1	
2b	$x^2 + 4x + 4x + 16$	M1	
	$x^2 + 8x + 16$	A1	
3a	$3(-2) + 2(5)$	M1	
	4	A1	
3b	$(4)^2 + (6)^2$	M1	
	52	A1	
3c	$200f + 50e$	B1	
3d	$200(20) + 50(120)$	M1	
	£10 000	A1	
4	$x^2 + ax + 12 = x^2 + (b + c)x + bc$	M1	
	Any that work eg $a = 7$, $b = 3$ and $c = 4$	A1	

175   Factorising quadratic expressions

Q	Answer	Mark	Comment
1a	$x(x - 4)$	B1	
1bi	$(x + a)(x + b)$ where $ab = 12$	M1	
	$(x - 6)(x + 2)$	A1	
1bii	6 and −2	B1	
2a	$x^2 - y^2$	B1	
2b	$(x + 7)(x - 7)$	B1	
2c	$(2a - 9)(2a + 9)$ or $(4a - 3)(4a + 3)$	M1	
	$(2a - 3)(2a + 3)$	A1	
3a	$6x^2 + 9x - 2x - 3$	M1	
	$6x^2 + 7x - 3$	A1	
3b	$(ax + c)(bx + d)$ where $ab = 6$ and $cd = 12$	M1	
	$(3x - 4)(2x - 3)$	A1	
4a	$(x + a)(x + b) = 0$ where $ab = 10$	M1	
	$(x + 5)(x - 2) = 0$	A1	
	−5 or 2	A1	
4b	$(x + a)(x + b) = 0$ where $ab = 5$	M1	
	$(x + 5)(x - 1) = 0$	A1	
	−5 or 1	A1	
5a	$4x^2 - 25$	B1	
5bi	15 and 25	B1	
5bii	45 and 55	B1	

176   Solving quadratic equations

Q	Answer	Mark	Comment
1a	$(ax + c)(bx + d)$ where $ab = 12$ and $cd = 1$	M1	
	$(3x + 1)(4x + 1)$	A1	
1b	$x = -\dfrac{1}{3}$; $x = -\dfrac{1}{4}$	B1	
2a	$x^2 = 4.5$	M1	
	± 2.12	A1	
2b	$4x(x - 5)$	M1	
	0 and 5	A1	

3a	$\dfrac{-(3) \pm \sqrt{((3)^2 - 4(1)(-7))}}{2(1)}$	**M1**	
	$\dfrac{-3 \pm \sqrt{37}}{2}$	**A1**	
	1.54, −4.54	**A1**	
3b	$\dfrac{-(5) \pm \sqrt{((5)^2 - 4(2)(-4))}}{2(2)}$	**M1**	
	$\dfrac{-5 \pm \sqrt{57}}{4}$	**A1**	
	0.64, −3.14	**A1**	
4a	$(ax + c)(bx + d)$ where $ab = 9$ and $cd = 2$	**M1**	
	$(3x - 1)(3x - 2) = 0$	**A1**	
	$x = \frac{1}{3}$ or $\frac{2}{3}$	**A1**	
4b	$(ax + c)(bx + d)$ where $ab = 8$ and $cd = 5$	**M1**	
	$(2x - 1)(4x + 5) = 0$	**A1**	
	$x = \frac{1}{2}$ or $-\frac{5}{4}$	**A1**	
5	$x(x - 6) = 1360$	**M1**	
	$x^2 - 6x - 1360 = 0$	**A1**	
	$(x + 34)(x - 40) = 0$	**M1**	
	40	**A1**	

177 Completing the square

Q	Answer	Mark	Comment
1a	$a = 4$; $b = 23$	**B2**	
1b	$x + 4 = \pm\sqrt{23}$	**M1**	
	$x = -4 \pm\sqrt{23}$	**A1**	
2a	$a = 5$; $b = -28$	**B2**	
2b	$x = -5 \pm \sqrt{28}$	**M1**	
	0.29, −10.29	**A1**	
3a	$a = 1$; $b = 6$	**B2**	
3b	Cannot take the square root of a negative number	**E1**	
4a	$a = -4$; $b = 14$	**B2**	
4b	$x = 4 \pm \sqrt{14}$	**M1**	
	7.74, 0.26	**A1**	
5	$p = -10$	**B1**	
	$q = 10$	**B1**	

178 Solving problems with quadratic equations

Q	Answer	Mark	Comment
1	$\dfrac{-(5) \pm \sqrt{((5)^2 - 4(1)(-9))}}{2(1)}$	**M1**	
	$\dfrac{-5 \pm \sqrt{61}}{2}$	**A1**	
2a	$(6x + 1)^2 = (5x + 4)^2 + (2x - 1)^2$	**M1**	
	$36x^2 + 12x + 1 = 25x^2 + 40x + 16 + 4x^2 - 4x + 1$	**A1**	It is sufficient to get to this stage as the next step is the answer that is given.
2b	$(ax + c)(bx + d)$ where $ab = 7$ and $cd = 16$	**M1**	
	$(7x + 4)(x - 4) = 0$	**A1**	
	4	**A1**	
3a	$x + \dfrac{1}{x} = 2.9$	**M1**	
	$10x^2 - 29x + 10 = 0$	**A1**	
3b	$(ax + c)(bx + d)$ where $ab = 10$ and $cd = 10$	**M1**	
	$(2x - 5)(5x - 2)$	**A1**	
	$\frac{2}{5}$ or $\frac{5}{2}$	**A1**	
4	$183.6 \div 19.2 = 9.5625$	**M1**	
	$x(x - 2) = 9.5625$	**M1**	
	$x^2 - 2x - 9.5625 = 0$	**M1**	
	$x = 4.25$ m	**A1**	

Workbook answers

179   Real-life graphs

Q	Answer	Mark	Comment
1ai	2	B1	
1aii	5	B1	
1bi	40	B1	
1bii	5 km in 40 mins	M1	
	7.5	A1	
2		B3	You will get a mark for each straight line.
3a	Line from (10, 0) to (12.30, 30)	B1	
	Line from (11, 0) to (11.30, 10) to (11.45, 10) to (12.45, 30)	B1	
3b	Jogger by 15 min	B1	

180   Linear graphs

Q	Answer	Mark	Comment
1		B2	You will get 1 mark for a line passing through (0, −1) and 1 mark for a line from (−3, −7) to (3, 5).
2a	Line A; 3; Line B: $\frac{1}{2}$; Line C: $-\frac{5}{3}$	B3	B1 for each correct answer.
2b		B3	B1 for each correct answer.
3a	Line from (0, 20) to (4, 140)	B1	
	Line from (0, 30) to (4, 130)	B1	
3b	Same at 2 hours	A1	
	Bernice as she is cheaper from 2 to 4 hours	E1	

181   Linear graphs

Q	Answer	Mark	Comment
1a	D	B1	
1b	C	B1	
1c	D and E	B1	
1d	Graph intersecting y-axis at −2	B1	
	Graph with gradient of 3	B1	
2a	Graph from (2, 0) to (0, −5)	B1	
2b	Graph from (3, 0) to (0, −2)	B1	
3ai	$y = x + 4$	B1	
3aii	$x = -3$	B1	
3bi	−7	B1	
3bii	$-\frac{1}{7}$	B1	

Workbook answers

182 Equations of lines

Q	Answer	Mark	Comment
1a	$y = \frac{1}{3}x - 1$	B2	B1 for $y = \frac{1}{3}x$.
1b	Line through (0, 4)	B1	
	Line with gradient 2	B1	
1c	(–3, –2)	B1	
2ab		B3	B1 for a point plotted at (200, 35). B1 for a point plotted at (150, 30). B1 for straight line joining the points and extending to the limits of the graph.
2c	$C = 15 + \frac{1}{10}n$	B2	B1 for $\frac{1}{10}n$.
3	$a = 1, b = 0, c = \frac{3}{5}, d = 2, e = \frac{1}{5}, f = 6$	B3	You will lose a mark for every error.

183 Linear graphs and equations

Q	Answer	Mark	Comment
1ab	Line from (3, 0) to (0, 2)	B2	B1 for line passing through either of these points.
	Line passing through (0, –2)	B1	
	Line with gradient 2	B1	
1c	$x = 1\frac{1}{2}, y = 1$	B1	
2a	$(\frac{1}{2}, -\frac{1}{2})$	B1	
2b	$\frac{1}{3}$	B1	
2c	Gradient –3	M1	
	$y = -3x + 1$	A1	
3a	Pass through (0, –3)	B1	
3b	Same gradient of $-\frac{1}{4}$	B1	
3c	$17x = 17$	M1	
	(1, 1)	A1	

184 Quadratic graphs

Q	Answer	Mark	Comment
1a	0, 1, 4, 9	B1	
1b		B2	You will get 1 mark for plotting the points and 1 mark for a smooth curve.

Workbook answers

1c	−1.45, 3.45	B1	Acceptable range −1.4 to −1.5, 3.4 to 3.5.
1d	$x = 1$	B1	
2a	7, −2, −1, 2	B2	You will lose 1 mark for each error.
2b		B2	You will get 1 mark for plotting the points and 1 mark for a smooth curve.
2c	−2.9, 0.9	B1	
2d	$x = −2.4$ and $x = 0.4$	B1	
3a	Pass through (0, −3)	B1	
3b	Same shape	B1	
3c	$x^2 = 4$	M1	
	2 and −2	A1	

185 Quadratic graphs

Q	Answer	Mark	Comment
1a	$x = −2$; $x = 3$	B1	
1b	$x = −3$; $x = 4$	B1	
1c	$(x^2 − x − 6) − (x^2 − 2x − 8)$	M1	
	$y = x + 2$	A1	
		A1	
	−2 and 4		
2a	−3.5, 0, 3.5	B1	
2b		M1	
	$(12x − x^3) − (11x − x^3)$		
	$y = x$	A1	
	−3.3, 0, 3.3	A1	

Workbook answers

Q	Answer	Mark	Comment
3a	Cannot draw a parabola through these points or x-coordinate of the vertex must be midway between the roots	B2	If you plot the points you will get a mark and giving a reason gets the other mark.
3b	$(x + 1)(x - 2) = 0$	M1	This has a vertex at $(\frac{1}{2}, -2\frac{1}{4})$.
	$x^2 - x - 2 = 0$	A1	

186 Other graphs

Q	Answer	Mark	Comment
1a	0.125, 0.0625	B2	B1 for each
1b		B2	1 mark for plotting points and 1 mark for a smooth curve.
1c	0.3	B1	
2	A $(-4, 0)$; B $(-2, 0)$; C $(4, 0)$; D $(0, -32)$	B4	B1 for each correct answer.
3	$2^{0.641} = 1.559$	B1	
	$1 \div 0.641 = 1.560$	B1	
	0.641	B1	

187 Algebraic fractions

Q	Answer	Mark	Comment
1a	$\frac{3(2x + 3) + 2(3x - 1)}{2 \times 3}$	M1	
	$\frac{12x + 7}{6}$	A1	
1b	$\frac{2(3x - 1) - 5(2x - 5)}{5 \times 2}$	M1	
	$\frac{-4x + 23}{10}$	A1	
1c	$\frac{(2x - 1)(4x - 1)}{3 \times 2}$	M1	
	$\frac{8x^2 - 6x + 1}{6}$	A1	
1d	$\frac{6(x - 1)}{3(3x - 1)}$	M1	
	$\frac{2x - 2}{3x - 1}$	A1	
2a	$\frac{2x + 3}{5} \times \frac{15}{6x + 9}$ or $\frac{15^3 (2x + 3)}{5^1 (6x + 9)}$	M1	Alternative methods are acceptable.
	$\frac{6x + 9}{6x + 9} = 1$	A1	
2b	$\frac{2x^2}{9} - \frac{6y^2}{9}$	M1	
	$\frac{2(x^2 - 3y^2)}{9}$	A1	
3	Numerator y^2	B1	
	Denominator y	B1	

188   Solving equations

Q	Answer	Mark	Comment
1a	$5(x + 1) + 2(x - 3)$	M1	
	$7x - 1$	A1	
	$7x - 1 = 20$	M1	
	3	A1	
1b	$5(4x + 1) - 3(3x - 1)$	M1	
	$11x + 8$	A1	
	$11x + 8 = 30$	M1	
	2	A1	
1c	$3(3x - 1) + 5(5x + 1)$	M1	
	$= 3(5x + 1)(3x - 1)$	M1	
	$34x + 2 = 45x^2 - 6x - 3$	A1	
	$45x^2 - 40x - 5 = 0; 9x^2 - 8x - 1 = 0$	M1	
	$(9x + 1)(x - 1) = 0$	A1	
	$x = -\frac{1}{9}$ or 1	A1	
2a	$3(3x - 1) - 4(2x - 1) = (2x - 1)(3x - 1)$	M1	
	$9x - 3 - 8x + 4 = 6x^2 - 5x + 1$	A1	
	$6x^2 - 6x = 0$	A1	It is sufficient to get to this stage as the next step is the answer which is given.
2b	$x(x - 1) = 0$	M1	
	0 or 1	A1	
3	$x(x + 1) + 3(x - 1)$	M1	
	$= (x + 1)(x - 1)$	M1	
	$x^2 + x + 3x - 3 = x^2 - 1$	A1	
	$4x - 2 = 0$	M1	
	$\frac{1}{2}$	A1	
4	Multiply top and bottom by $2x + 1$	M1	
	Numerator $6x^2 + x - 1$	B1	
	Denominator $4x^2 - 1$	B1	

189   Simultaneous equations 2

Q	Answer	Mark	Comment
1	$x^2 + (4 - x) = 16$	M1	
	$x^2 - x - 12 = 0$	M1	
	$(x - 4)(x + 3)$	A1	
	$(4, 0)$	A1	
	$(-3, 7)$	A1	
2a	$y = (3x - 13)/2$	B1	
2b	$x^2 + ((3x - 13)/2)^2 = 65$	M1	
	$4x^2 + 9x^2 - 78x + 169 = 260$	A1	
	$13x^2 - 78x - 91 = 0; x^2 - 6x - 7 = 0$	M1	
	$(x - 7)(x + 1) = 0$	A1	
	$A = (7, 4)$	A1	
	$B = (-1, -8)$	A1	
3	$9y^2 + 2y^2 = 11$	M1	
	$y^2 = 1$	A1	
	$(3, 1)$	A1	
	$(-3, -1)$	A1	
4a	$(5 - 2x)^2 + x^2 = 5$	M1	
	$25 - 20x + 4x^2 + x^2 = 5$	A1	
	$5x^2 - 20x + 20 = 0; x^2 - 4x + 4 = 0$	M1	
	$(x - 2)(x - 2) = 0$	A1	
	$x = 2, y = 1$	A1	
4b	1 as there is only 1 solution	B1	

Workbook answers

190 The nth term

Q	Answer	Mark	Comment
1a	5, 9, 13	B1	
1b	7th	B1	
1c	not odd	B1	
2	$7n - 4$	B2	You will get 1 mark for writing down $7n$.
3a	Always odd	B1	
3b	Always even	B1	
3c	Could be either	B1	
3d	Always odd	B1	
3e	Always odd	B1	
4a	$2 \times$ anything is even	B1	
4b	$2n \times 2n$	M1	
	$4n^2$ is a multiple of 4	A1	
5	A: 5, 7, 9, 11, 13, 15, 17, 19, … and B: 4, 9, 14, 19, ….	M1	
	9, 19, 29, 39, ..	A1	
	$10n - 1$	A1	

191 Formulae

Q	Answer	Mark	Comment
1a	16, 21, 26	B1	
1b	101	B1	
1c	$5n + 1$	B2	You will get 1 mark for $5n$.
2	$y(x - 4) = x + 2$	M1	
	$yx - 4y = x + 2$	A1	
	$yx - x = 2 + 4y; x(y - 1) = 2 + 4y$	M1	
	$x = \dfrac{2 + 4y}{y - 1}$	A1	
3	$6x + 2y = 4x + 12$	M1	
	$2x = 12 - 2y$	A1	
	$x = 6 - y$	A1	
4	$2xy - x = 1 + y$	M1	Multiply by denominator
	$x = (1 + y)/(2y - 1)$	A1	
	$z = \dfrac{\frac{1+y}{2y-1} + 3}{\frac{2(1+y)}{2y-1} - 1}.$	M1	
	$= \dfrac{1 + y + 3(2y - 1)}{2(1 + y) - (2y - 1)}$	A1	
	$z = \dfrac{1 + y + 6y - 3}{2 + 2y - 2y + 1}$	A1	It is sufficient to get to this stage as the next step is the answer which is given.

192 Inequalities

Q	Answer	Mark	Comment
1a	$3x \leqslant 6$	M1	
	$x \leqslant 2$	A1	
1b	$x > -2$	B1	
1c	$-1, 0, 1, 2$	B1	
2a	$-3 < x \leqslant 1$	B1	
2bi	$\dfrac{x}{2} > -2$	M1	
	$x > -4$	A1	
2bii	$x + 3 \leqslant 2$	M1	
	$x \leqslant -1$	A1	
2c	$-3, -2, -1$	B1	
3a	$3x - x \geqslant 7 + 2$	M1	
	$2x \geqslant 9$	A1	
	$x \geqslant 4.5$	A1	
3b	$3x - 3 < x - 5$	M1	
	$2x < -2$	A1	
	$x < -1$	A1	
4	3, 5, 7, 9	B2	You will lose a mark for every missing or incorrect value.

Workbook answers

193 Graphical inequalities

Q	Answer	Mark	Comment
1	$x \geqslant -4$	B1	
	$y \leqslant 3$	B1	
	$y \geqslant x$	B1	
2		B3	You will get 1 mark for each line and lose a mark if you do not mark R clearly.
3	M: b, c, g, h F: a, d, e T: f	B3	You will lose 1 mark for each wrong answer.

194 Graph transforms

Q	Answer	Mark	Comment
1a		B1	
1b		B1	
1c		B1	
2	360, 720 etc	B1	
3	Sketch of both graphs	B2	B1 for each graph
	2	B1	

195   Graph transforms

Q	Answer	Mark	Comment
1a		B1	
1b		B1	
1c		B1	
2a	$(x + 1)^2 - 7$	B2	B1 for $(x + 1)^2$
2b	Translation $\binom{-1}{0}$	B1	
	Translation $\binom{0}{-7}$	B1	You can put these together as a single translation $\binom{-1}{-7}$.

196   Proof

Q	Answer	Mark	Comment
1	$n^2 + 10n + 25 - (n^2 + 6n + 9)$	M1	
	$= 4n + 16$	A1	You need to show that the minus sign outside the bracket changes the sign of each term inside the bracket.
	$= 4(n + 4)$	A1	
2	CDB = 75° (alternate segment) CDB = 35° (alternate segment, angles in a triangle) DAB = 110° (opposite angles in cyclic quad) ADB = ABD = 35° (alternate angles)	B2	You only need to give two of these but you must give a reason for each one you give.
	ADB = DBC (alternate angles) or ADC + DCB = 180° (angles on line)	B1	You only need one reason why the lines are parallel, but a reason must be given.
3	$8 \times \frac{1}{2}n(n + 1) + 1 = 4n(n + 1) + 1$	M1	
	$4n^2 + 4n + 1$	A1	
	$(2n + 1)^2$	A1	
4	$\overrightarrow{OX} = \overrightarrow{OM} + \overrightarrow{MX} = \mathbf{a} + \frac{1}{3}(-\mathbf{a} + 2\mathbf{b})$	M1	
	$= \frac{2}{3}\mathbf{a} + \frac{2}{3}\mathbf{b}$	A1	
	$\overrightarrow{OX} = \overrightarrow{ON} + \overrightarrow{NX}$		
	$= \mathbf{b} + \frac{1}{3}(-\mathbf{b} + 2\mathbf{a})$	M1	
	$= \frac{2}{3}\mathbf{a} + \frac{2}{3}\mathbf{b}$	A1	

Notes

Student notes

Student notes